What is statistics?

Probability and statistics often go hand in hand. The way to consider the two is as follows:

Suppose there is a box containing 100 multi-colored ping pong balls. In a probability class, you would be told about the contents of the box:

50 red
30 white
20 blue

You would then be asked to determine the **probability** of pulling out a particular color combination if you randomly selected a certain number of balls from the box.

If you select 10 ping pong balls at random, what is the probability you will select 5 red, 3 white and 2 blue?

In a statistics class, you would not be told what about the contents of the box; rather, you would be told about the sample you have selected. You then would use that information to make inferences about what is in the box.

In a random sample of ten ping pong balls, four were blue. Find a 95% confidence interval for the proportion of all ping pong balls that are blue.

While the notion of a confidence interval may not make sense at this point, the overall meaning is that you would be asked to determine what proportion of the ping pong balls in the original box were blue. Obviously there is no way you would be able to determine exactly what proportion of the box consisted of blue ping pong balls (unless you counted them all,) but using **statistics** you can obtain a good estimate.

Proportions

A proportion is a decimal between 0 and 1. It is essentially the decimal equivalent of a percentage. It can be presented three ways in class:

Decimal form

Percent form

"Out of" statement form

Here is an example of the same situation expressed in those three different ways. In each instance, we want you to find the proportion from the sample.

In a sample of 200 VCU students...

Decimal: **...the proportion who vote is .7** In this case, the sample proportion would simply be .7

Percent: **...70% vote.** Here, to convert the percent to a decimal, simply slide the point two places to the left. The sample proportion works out to be .7

Out of Statement: **...140 vote.** Although it doesn't say "out of" expressly, you can re-word the statement to say "140 out of 200" vote. "Out of" in math means divide, so the math problem becomes 140/200 which equals .7. That is your sample proportion. (If you divide in the wrong order, you'll know the answer isn't right because you'll come out with an answer larger than one.)

Determine the following sample proportions. Round to the nearest ten-thousandth when appropriate (four decimal places).

1. In a random sample of 500 people, 260 say they give to charity.

2. In a random sample of 40 people, 10% smoke.

3. 52 out of 300 people have traveled out of the country.

4. The proportion of people who have been in car accidents is .364

5. 78.3% of people have flown in airplanes.

6. A random sample of 368 VCU students revealed that 57 were graduate students.

Experiments vs. Observational Studies

Once you have a sample, there are two different types of studies you can do:

Experiment

Observational Study

Example of an experiment:

A group of people suffering from back pain agrees to try a new medication, designed to reduce the pain. The researcher randomly chooses half the people to take the real drug, and the other half gets the placebo. At the end of six months, there appears to be a relationship (or association) between taking the drug and reduced back pain. You may conclude the pill caused the reduction in back pain.

Example of an observational study:

You wish to see if men or women are better drivers. You amass a group of people and determine how many points they have on their licenses. You average the male points and the female points and compare to see if one number is higher than the other. If at the end it appears women have fewer points, on average, than men do, you may conclude there is a relationship/association between gender and driving ability. You may NOT conclude that being female causes people to be better drivers.

The back pain example is an experiment because

1) The researcher assigned the groups, and
2) There was a behavior change (people are now taking a daily pill that they weren't before)

DO NOT use the fact that cause and effect was declared as a means for concluding this was an experiment. Sometimes journalists don't know the rule about cause and effect, and they draw that conclusion at the end of observational studies.

The driving example was an observational study because

1) The groups (male/female) came already formed. The researcher did not assign people to be a particular gender
2) There was no behavior change. Their driving record was simply obtained.

Determine if the following are experiments or observational studies. Be prepared to explain why:

1. A group of 150 students was randomly selected. A third of them were randomly selected to watch a video produced by Democrats, a third were told to watch a video produced by Republicans, and the last third watched nothing. A survey was given to the students afterwards, and differences in opinion were noted.

2. A group of 50 people who regularly do yoga was randomly selected, and a group of 50 people who have never done yoga was randomly selected. Their resting pulse rates were measured, and the difference in average resting pulse rate was determined.

3. A manual dexterity test was given to people, and the speed with which they could complete it was noted. The average speed of left-handed people was compared to the average speed of right-handed people.

4. To test the effectiveness of a new allergy pill, 1000 allergy sufferers were randomly divided into two groups; one group took the pill, one group took a placebo. After a month, allergy symptoms were measured and compared.

Vocabulary

Explanatory variable:

Response variable:

Refer to the previous page, stating the explanatory and response variables for each study.

Individuals:

Population:

Sampling Frame:

Sample:

Variable:

Researchers wanted to see if a new stretching exercise could reduce back pain in those whose back pain was considered moderate. 170 moderate back pain sufferers were randomly selected from the patient lists of chiropractors state-wide. 85 randomly selected patients from that group were told to do the daily stretching exercise in addition to their current treatment. The remaining 85 simply stuck with their current treatment.

After six months, the group who had done the stretching exercises reported significantly less pain than the group who hadn't.

What is the population of interest?

What is the sampling frame?

What is the sample?

Is this an experiment or an observational study? Why?

What was the response variable?

Can you declare that stretching caused the reduction in back pain? Explain.

You wish to see the quality of the flights that are coming into Richmond Airport on a particular airline. You stand by the baggage claim area for every flight from that airline on a Tuesday, interviewing people who are waiting for their luggage, asking them if they enjoyed their flight.

What is the population of interest?

What is the sampling frame?

Is this an experiment or an observational study? Why?

There is no explanatory variable here, but what is the response variable?

Can you take the results from the sample and apply them to the whole population?

Sample Collection Methods

If you were to do a **census**, what would you do?

Simple Random:

Convenience:

Voluntary Response:

Systematic:

Cluster:	Stratified:	Multistage:

Stratified

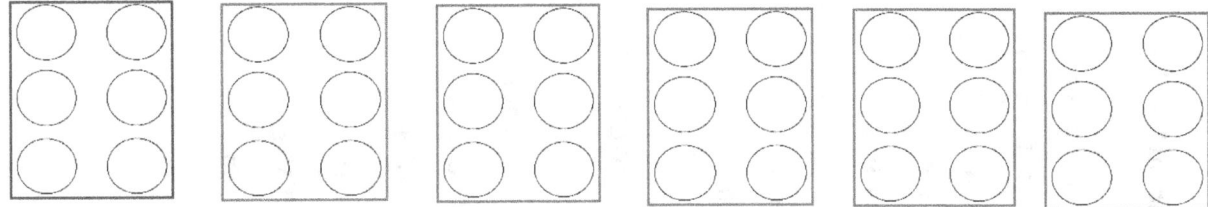

Cluster

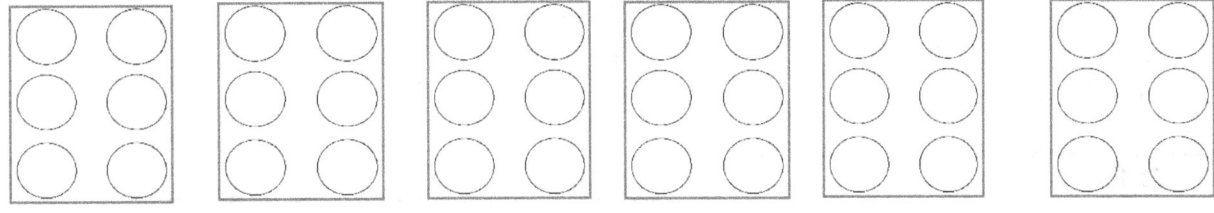

Multistage

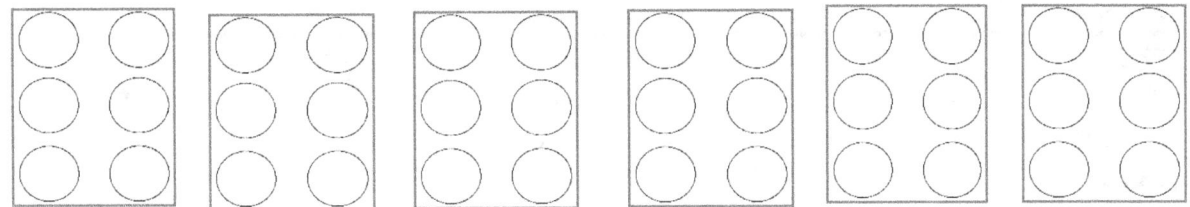

Student Number	Lab Instructor	Simple Random	Stratified	Cluster	Multistage	Convenience	Census	Systematic	Student Number	Lab Instructor	Simple Random	Stratified	Cluster	Multistage	Convenience	Census	Systematic
1	Smith								32	Walker							
2	Smith								33	Walker							
3	Smith								34	Walker							
4	Smith								35	Walker							
5	Smith								36	Walker							
6	Smith								37	Walker							
7	Smith								38	Walker							
8	Smith								39	Walker							
9	Smith								40	Walker							
10	Smith								41	Walker							
11	Jones								42	Walker							
12	Jones								43	Johnson							
13	Jones								44	Johnson							
14	Jones								45	Johnson							
15	Jones								46	Johnson							
16	Jones								47	Johnson							
17	Jones								48	Johnson							
18	Jones								49	Johnson							
19	Jones								50	Johnson							
20	Jones								51	Johnson							
21	Jones								52	Johnson							
22	Bauer								53	Turner							
23	Bauer								54	Turner							
24	Bauer								55	Turner							
25	Bauer								56	Turner							
26	Bauer								57	Turner							
27	Bauer								58	Turner							
28	Bauer								59	Turner							
29	Bauer								60	Turner							
30	Bauer								61	Turner							
31	Bauer								62	Turner							

List the name of the sample collection method described:

You wish to see passenger satisfaction for all airline passengers who come through Richmond International Airport. You select your sample by...

Standing at a booth with a sign, inviting people to tell you about their experience.

Selecting every 20th passenger that walks by.

Randomly selecting three flights, and then using every passenger from those three flights.

Using three people from every flight.

Using every passenger from every flight.

Randomly selecting ten flights, and then randomly selecting eight passengers from each of those ten flights.

Obtaining a list of passengers, using software to determine which passengers to choose

Use the first 50 people you see

Vocabulary

Parameter:

Statistic:

Bias:

Variability:

You wish to see what proportion of all teens have tried smoking. In order to conduct the study, you go to an urban high school and randomly select 100 students to interview. You call them into an office, and, in front of their parents, ask if they've tried smoking. 8% of the students admitted to smoking.

You repeat the procedure at rural and suburban high schools. The results are 7% and 8%, respectively.

What is the population of interest?

What is the parameter?

What are the statistics?

Does bias exist?

Would increasing the sample size eliminate the bias?

Does variability exist?

You wish to see if a particular class is worth taking. There are twenty-seven sections of this class available, and you interview several random students from ten randomly selected sections, asking them if they recommend the class. For the ten sections, the following percent of students said they recommend it.

10%   80%   37%   14%   92%   50%   72%   22%   65%   8%

What is the population of interest?

What is the parameter?

What are the statistics?

Does bias exist?

Does variability exist?

Would increasing the sample size help with the variability?

You wish to see what proportion of football fans believe the AFC will win the Superbowl this year. You obtain a sample of 400 randomly selected football fans, and 52% of them say yes, the AFC will win the Superbowl.

What is the population of interest?

What is the parameter?

What is the statistic?

Can we say conclusively that, since 52% of our sample said yes, that exactly 52% of the population will say yes?

Since we got our sample randomly, can we say the proportion of the population who says yes should be around 52%?

If we got our sample from Denver, the home of an AFC team, could we say that somewhere around 52% of all fans believe the AFC will win?

What do we mean by somewhere around 52%?

Confidence Intervals

Margin of error:

Formula for margin of error:

How to create a confidence interval and make a confidence statement:

1. Find p-hat (can be given as decimal, %, or "out of" statement)
2. Find margin of error by doing 1/(sqrt n)   where n = sample size
3. Add/subt margin of error to/from the p-hat value (be sure to be consistent with units)
4. Make confidence statement

Three parts to a confidence statement:

1. State how confident you are (always 95% for now)
2. State the interval
3. Tie statement to the problem, referencing all or population

Example:

You wish to see what proportion of football fans believe the AFC will win the Superbowl this year. You obtain a sample of 400 randomly selected football fans, and 52% of them say yes, the AFC will win the Superbowl.

Find the sample proportion $\hat{p}$:

Calculate the margin of error:

Compute the interval:

Make the confidence statement:

A random sample of 625 people revealed that 375 have used mass transit in the past year. Calculate and interpret a 95% confidence interval for the proportion of all people who have used mass transit in the past year.

Find the sample proportion $\hat{p}$:

Calculate the margin of error:

Compute the interval:

Make the confidence statement:

Flip	Heads?	Flip	Heads?	Flip	Heads?	Flip	Heads?	Flip	Heads?
1		6		11		16		21	
2		7		12		17		22	
3		8		13		18		23	
4		9		14		19		24	
5		10		15		20		25	

Activity:

We wish to determine what proportion of coin flips land heads up. To conduct the study, flip your coin 25 times, marking a check in the "Heads?" column if it lands with heads up.

Determine $\hat{p}$ from your sample:

Is $\hat{p}$ going to be the same for everybody?

Calculate the margin of error:

Determine your interval:

Make your confidence statement:

What does a 95% confidence interval mean? It means 95% of your intervals should contain the correct population value of .50

| 0 | .10 | .20 | .30 | .40 | .50 | .60 | .70 | .80 | .90 | 1 |

Questions to Ask When Reading About a Study

1. **Who funded and conducted the study?**

   Is it an objective source looking for the truth, or is it a company trying to sell product? Is it someone with a point to prove, perhaps stretching the truth?

2. **Who were the individuals or objects studied and how were they selected?**

   Was a good, random sampling method used? Does the sample represent the population?

3. **What was the setting in which the study was done?**

   Anything done in a lab is suspect.

4. **What was the exact nature of the measurements made or questions asked?**

   How were certain variables measured?

5. **Are there any other differences in the groups being compared?**

   In observational studies, this can really be a problem. Suppose all of the people in one of the groups have something else in common (which could explain the outcome?)

6. **What was the magnitude of any claimed effects or differences?**

   Were numbers given, or phrases like "twice as effective?"

A Nicotine Patch Study

The people at "Nicopatch" (a fictitious company that makes nicotine patches) conducted a study to determine the effectiveness of their new patches. In the week-long study conducted in a residential facility, patches were found to be three times more effective in getting people to quit smoking than when going cold turkey.

Who funded/conducted the study? Are they trustworthy?

Who were the individuals studied and how were they selected?

What was the setting in which this study was done?

How were the variables (quitting) measured?

Were there any other differences in the groups being compared?

What was the magnitude of any claimed effects or differences?

Can cause and effect be claimed?

Exercise Lowers Blood Pressure

A recent study conducted by a national gym chain determined that exercising ten or more hours per week causes lower blood pressure. A group of people who worked out ten or more hours per week was compared to a group who didn't exercise at all, and the average blood pressure for the exercise group was significantly lower than those who didn't.

Who funded/conducted the study? Are they trustworthy?

Who were the individuals studied and how were they selected?

What was the setting in which this study was done?

How were the variables measured?

Were there any other differences in the groups being compared?

What was the magnitude of any claimed effects or differences?

Can cause and effect be claimed?

Mattress Improves Sleep

An independent research firm determined that sleep improves on a sturdy-coil mattress. Three-hundred randomly selected participants had their sleep measured on their current beds in their homes, and then had their sleep re-measured after switching to a sturdy-coil mattress. 87% of people had an increase in REM sleep when they slept on a sturdy-coil mattress when compared to their regular mattress.

Who funded/conducted the study? Are they trustworthy?

Who were the individuals studied and how were they selected?

What was the setting in which this study was done?

How were the variables measured?

Were there any other differences in the groups being compared?

What was the magnitude of any claimed effects or differences?

Can cause and effect be claimed?

More questions to ask when reading about a study

1. Is any information missing?

2. What is the sample?

3. Is the math correct?

   a. When do percentages need to add to make 100%?

   b. Should counts or percentages be used?

4. Are the numbers reasonable?

5. Could a better explanation exist?

Explain the issue with the following statements:

1. These supplements are twice as effective as the leading competitor.

2. People who eat breakfast weigh less than people who don't, so you should buy our breakfast cereal to lose weight.

3. 61% of people say they wouldn't recommend the competitor's product, so that means the majority of people would recommend ours.

4. 80% of people surveyed were vegetarian. Therefore, around 80% of people are vegetarian.

5. The average savings for people who switched to our company was $435. Therefore, our company will save you money.

6. 53% of people own a dog, so 47% own a cat.

7. 40 students from school A failed the SOL. 34 students from school B failed the SOL. School B does a better job preparing the students for the SOL.

Surveys

1. How old are you?

2. What is your gender?

3. Do you think the diameter of the moon is more or less than 1000 miles?

4. Compared to other drivers at VCU, would you rate yourself as better than average, average, or worse than average?

5. If you found a wallet with $20 in it, would you do the right thing and return the money?

6. Do you support the Greater Richmond Employment Act?

7. How far is the earth from the sun?

8. Do you eat chocolate and asparagus often?

9. What is the diameter of the moon?

Survey Vocabulary

Quantitative (numerical) data:

Qualitative (categorical) data:

Bias:

Deliberate bias:

Desire to please:

Asking the uninformed:

Ordering of questions:

Unnecessary complexity:

Confidential:

Anonymous:

33

Determine if the following are good questions

If they have a flaw, identify it.

Assume you will put your name on this survey and return it to me.

1. How many servings of vegetables have you eaten over the past few days?

2. Am I doing a good job teaching this class?

3. How much time should the physics department spend teaching the Dirac Equation?

4. How many credits are you taking this semester?

5. Are you physically fit?

6. How many pounds can you bench press?

7. You are planning to vote Democratic in the next election, right?

8. Are you married?

Design of Experiments

Placebo:

Control group:

Double blind:

Single blind:

Confounding variable:

Randomization:

Pairing or blocking:

Interacting variables:

Hawthorne effect:

Extending the results:

A study was recently done to see the effectiveness of a new sleep aid. A group of 100 volunteers were randomly selected to either receive a placebo or the experimental drug. To monitor the amount of time it took for the subjects to fall asleep, a person who was unaware of group assignment sat in a hospital room with the subject and clocked how long it took for sleep to arrive. Prior to taking the pill, subjects filled out a questionnaire about caffeine consumption, as that might interfere with the subject's ability to sleep.

What is the treatment?

Who comprises the control group?

Would confounding be an issue here? Why or why not?

Is this single or double blind?

What interacting variable has been considered?

Can the researchers extend the results? Why or why not?

Is this an experiment or an observational study?

A study was done to see if a manual dexterity test could be performed faster by left-handed or right-handed people. At the completion of the test, a buzzer sounded, and the person recording the times used that as in indication that the timer needed to be stopped. The person timing did not know if the subject was left or right handed. Subjects were asked how much sleep they had gotten the night before because that might impact the results.

Is this an experiment or an observational study?

Is it single or double blind?

What interacting variable was considered?

A study was done to see if taking a multivitamin daily would increase bone density. A random sample of 1000 people was selected; half were randomly assigned to take the vitamin, the other half took a placebo. After six months, the increase in bone densities between the groups was compared.

Can the results of that study be extended?

Can the researchers say that the multivitamin caused the increase in bone density?

A researcher wanted to see if putting a speed-monitoring device in cars would reduce the number of speeding tickets received. A group of 200 people were randomly selected and were allowed to choose whether they wanted the speed monitoring device or not. They were told that at the end of three months, the average number of speeding tickets for each group would be compared.

Give an example of a potential confounding variable.

What poor experimental method led to the possibility of confounding?

How could the Hawthorne Effect cause a problem here?

Who comprises the control group?

Ethics, Validity, Bias and Reliability

The three things that need to be in place before doing an ethical study on people:

Valid:

Rates vs. Counts:

Reliable:

Bias:

Determine if the following situations have reliability issues, validity issues, bias or none of these.

1. A large, inner-city hospital has more deaths than a small, rural hospital. Therefore, we can conclude the inner-city hospital is not as good as the rural hospital.

2. Temperatures are measured with a thermometer that is known to consistently run hot.

3. Temperatures are measured with an accurate thermometer.

4. Temperatures are measured with an inconsistent thermometer.

5. Viewers are asked to call in with their opinion.

6. When using a balance to weigh the same item, the weight comes out different each time.

7. High-pressure, standardized tests are used to determine intelligence.

Scatterplots

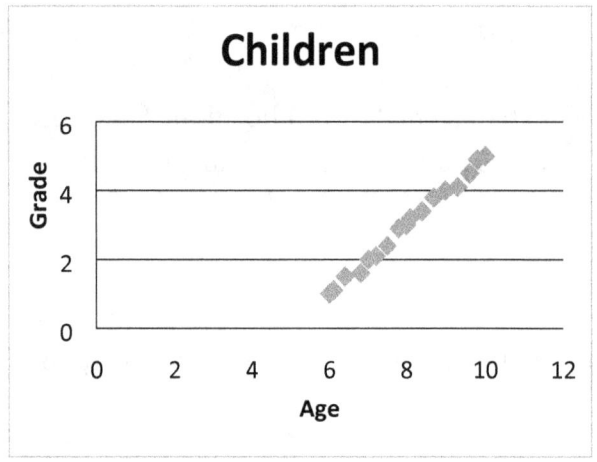 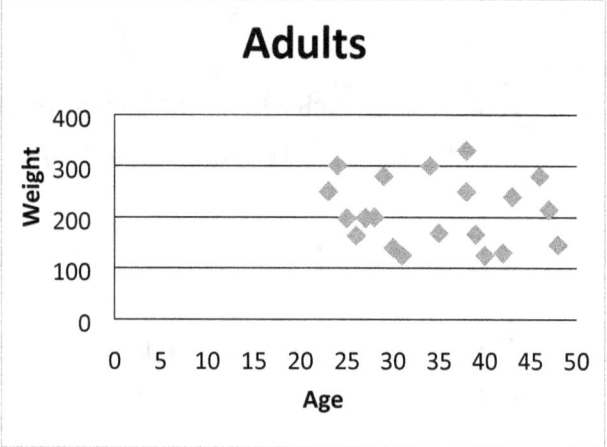

The **correlation coefficient**, r, measures how well the points from a straight line on a scale from 0 to 1. The sign indicates the direction.

The **coefficient of determination,** r^2, measures the percent of variation in y that is explained by x

How to create scatterplots on calculators:

TI 83/4

Hit [STAT] and [ENTER] to get your lists
Type all of the X-values into L1
Type all of the Y-values into L2
Hit [2nd] [Y=] to get to the STAT PLOT menu
Select the plot by hitting [ENTER]
Turn the plot on by hitting [ENTER]
Select the first option, scatterplot
Select L1 for Xlist (by hitting [2nd] [1])
Select L2 for Ylist (by hitting [2nd] [2])
Choose the mark you wish to use
Hit Graph
Zoom on the correct area by hitting [ZOOM] [9]

TI Nspire

From the home screen, press [MENU]
Choose option 4: Lists and Spreadsheets
Name your lists by clicking on the top cell and typing it in
Type in the values, hitting enter after each one
From the home screen, choose option 5, Data and Statistics, hit [ENTER]
Using the mouse in the arrow keys, scroll over the x and y axes and choose the correct variable for each

CASIO

Choose STAT from the home screen
Enter X values into List 1
Enter Y values into List 2
Select [F1] for graph 1

Create a scatterplot out of the following data set:

Hours Spent Studying	Grade on the Exam
0.7	65
1.1	71
1.3	73
1.5	78
2.7	82
2.8	87
3.4	92

The scatterplot should look as follows:

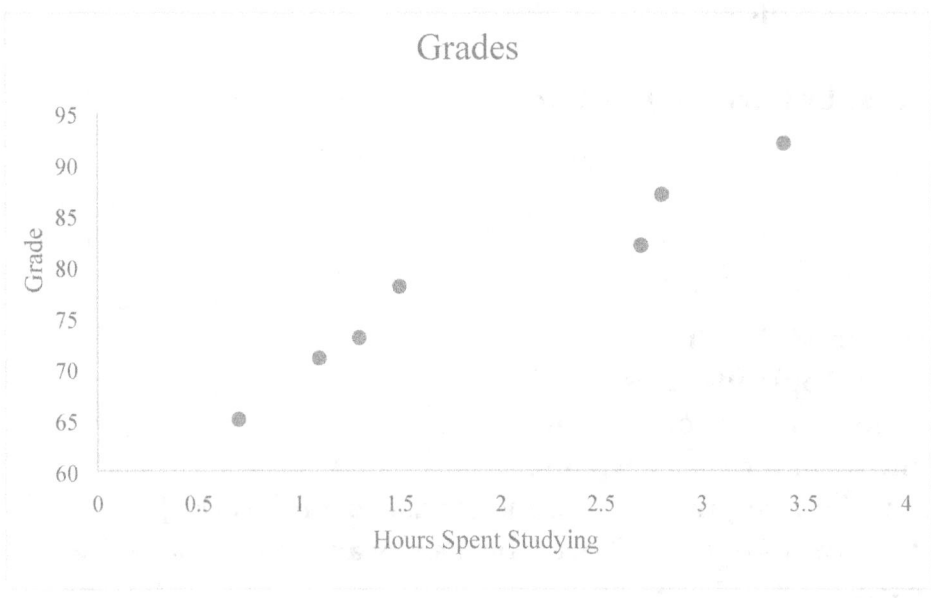

Correlation Coefficient:

Should this correlation coefficient be close to zero or one? Should it be positive or negative?

How to Determine Correlation Coefficient on Calculators:

TI 83

Turn the diagnostics on (this only has to be done once) by hitting [2nd] [0] for catalog, scrolling down to DiagnosticsOn and hitting [ENTER] [ENTER]

After entering the data into lists,
Hit [STAT] and arrow over to CALC
Select option 8: LinReg (a+bx)
Select appropriate lists by following the prompts or by separating them with a comma
r = the correlation coefficient

TI Nspire

After typing in the lists, find an empty cell and hit [MENU]
Select option 4: Statistics
Select option 1: Stat calculations
Select option 3: Linear Regression (mx+b)
Type in the names of the appropriate X and Y lists
Click OK
r is the correlation coefficient

Casio:

After typing in lists, select F1 for GPH1
Select F1 for CALC
Select F2 for X (linear)
r = the correlation coefficient

How to find the equation of the regression line

Create a scatterplot out of the following data set:

Hours Spent Studying	Grade on the Exam
.7	65
1.1	71
1.3	73
1.5	78
2.7	82
2.8	87
3.4	92

The scatterplot should look as follows:

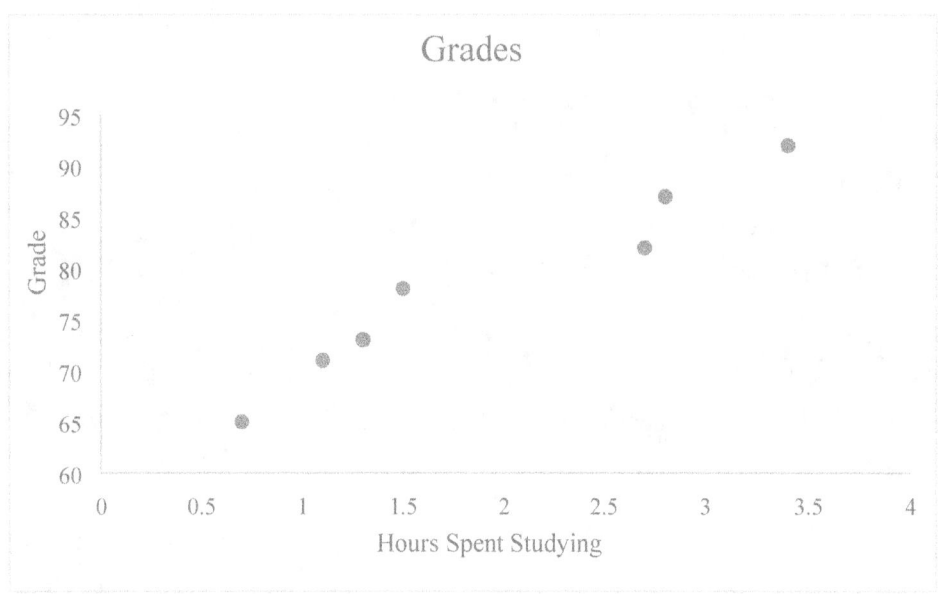

Correlation Coefficient value:

We wish to find the equation of the line that fits this data best. That way, we can use that line to predict future values.

The line that fits the points best is illustrated below.

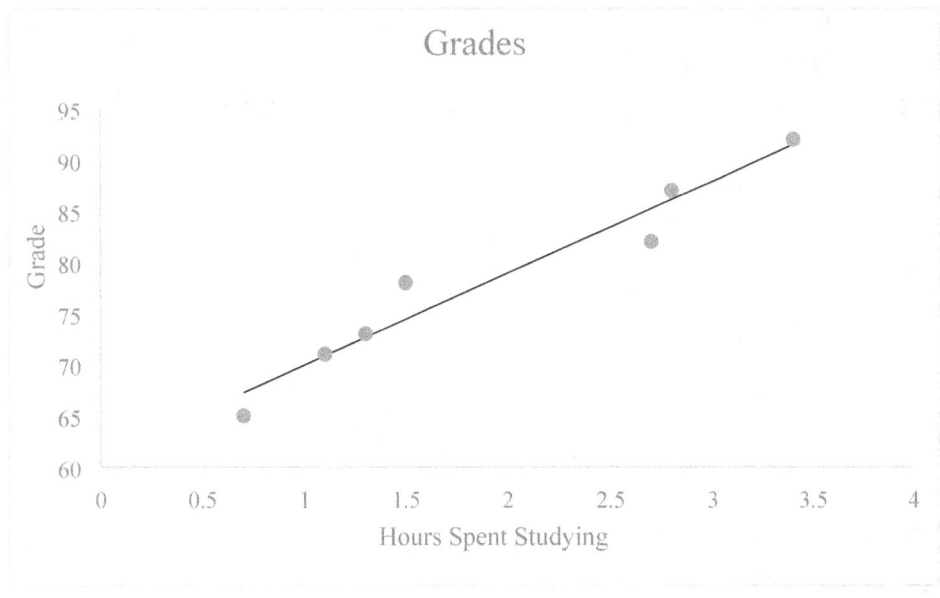

We can use the equation of that line to predict for missing values, such as 2.

What is the equation for that line (procedure is on next page)?

What would be the predicted grade if somebody spent 2 hours studying?

Finding the Equation for the Line of Best Fit

TI 83/84
After entering the data into lists,
Hit [STAT] and arrow over to CALC
Select option 8: LinReg (a+bx)
Select appropriate lists by following the prompts or by separating them with a comma
You are given the template on top, y = a+bx
You are given a and b to plug in.
Y and x stay as y and x

TI Nspire
After typing in the lists, find an empty cell and hit [MENU]
Select option 4: Statistics
Select option 1: Stat calculations
Select option 3: Linear Regression (mx+b)
Type in the names of the appropriate X and Y lists
Click OK
You are given the template y = mx + b
The calculator tells you m and b
X and Y stay x and y

Casio
After typing in lists, select F1 for GPH1
Select F1 for CALC
Select F2 for X (linear)
r = the correlation coefficient

The following table relates the number of hours spent training and the time to complete a race.

Hours Training	Minutes to Complete Race
27	21
24	23.6
9	32.1
15	29.6
28	21.8
30	22.4
21	25.9
17	28.4

What is the correlation coefficient for this data set?

What is the best equation to relate the training hours and race time?

Is this equation good for predicting race times based on training?

What would be the predicted finishing time of a person who trained for the race for 23 hours?

What would be the predicted finishing time for somebody who trained for 100 hours?

Extrapolation

Extrapolation:

The following table relates the horsepower of an engine and the gas mileage for small cars.

Horsepower	Miles per Gallon
180	30.4
190	29.8
190	30.1
200	28.6
210	28.1
210	27.9
220	27.6

What is the equation that relates horsepower with gas mileage?

Is this a good predictor?

What would be the expected miles per gallon for a car that has 205 horsepower?

Should this be used to determine the gas mileage of a pickup truck?

A pickup truck has 420 horsepower. According to this formula, what would be the predicted miles per gallon?

When is each graph appropriate?

Pie chart:

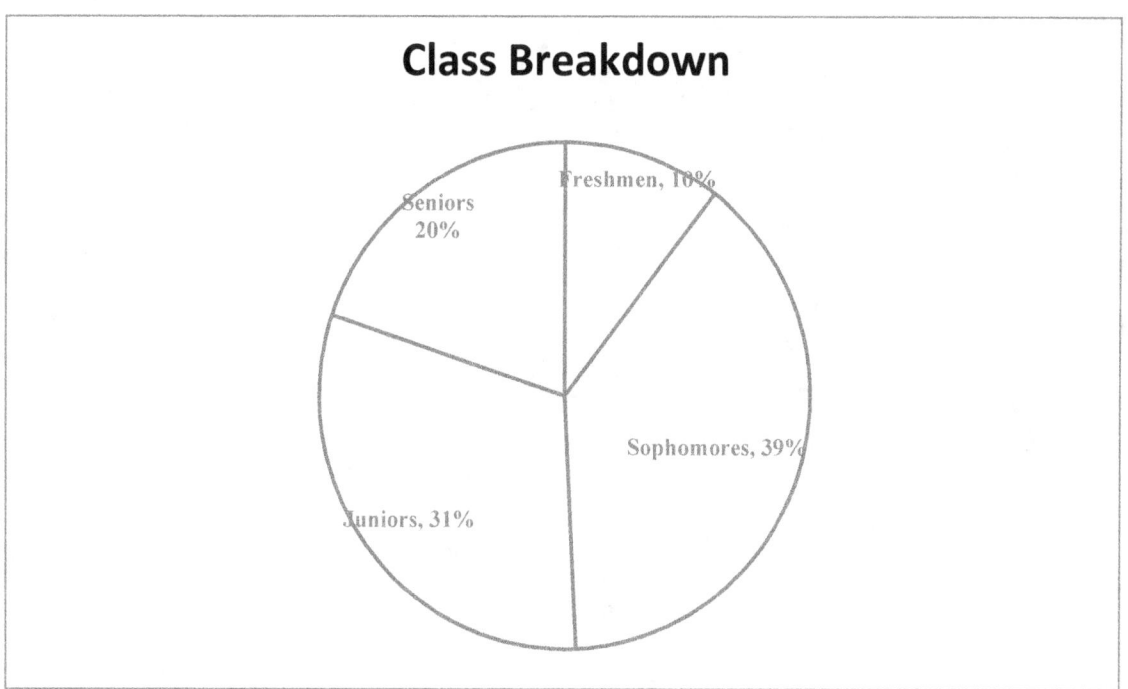

Bar Graph:

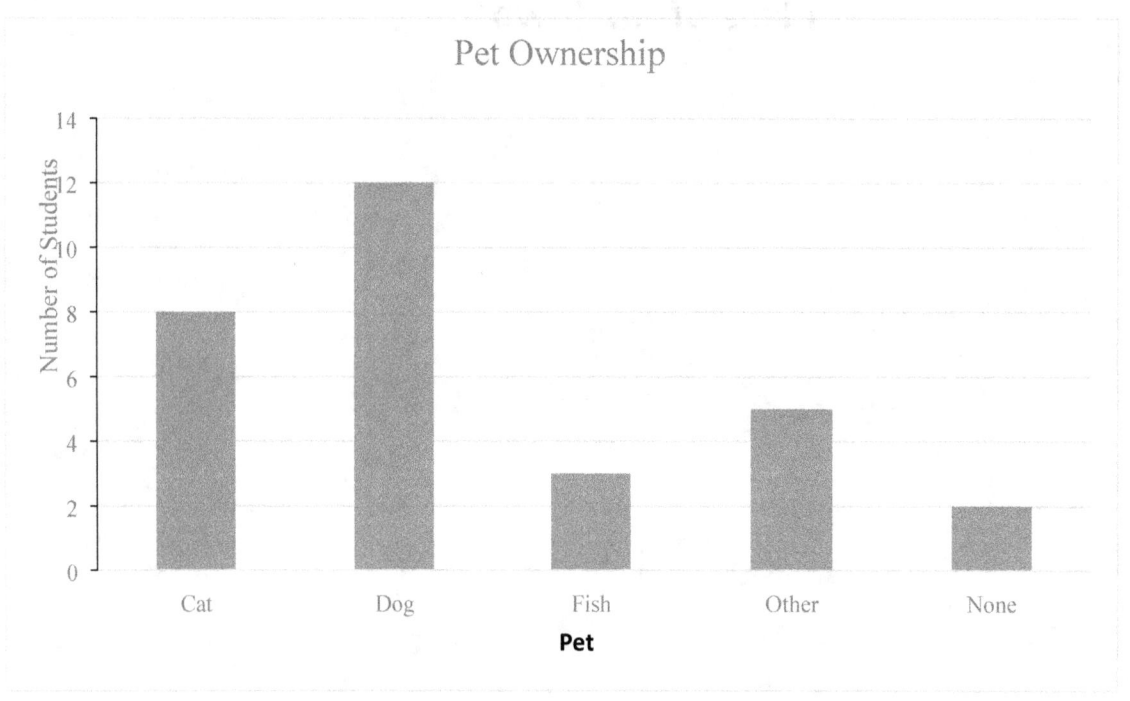

Line graph:

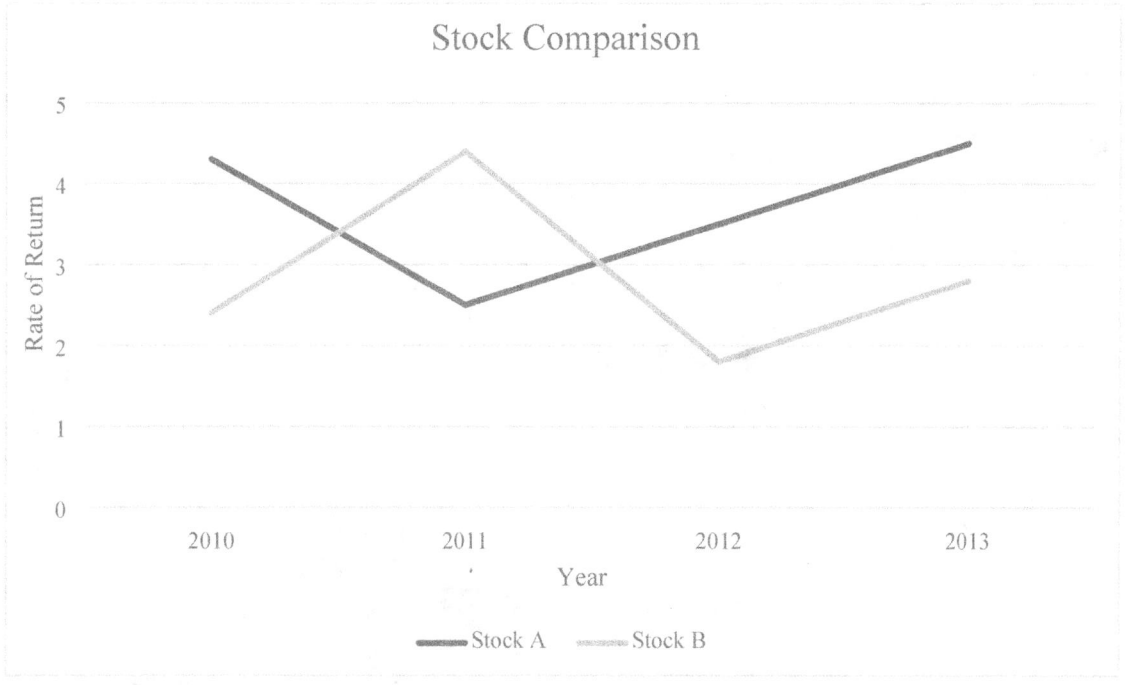

Histogram:

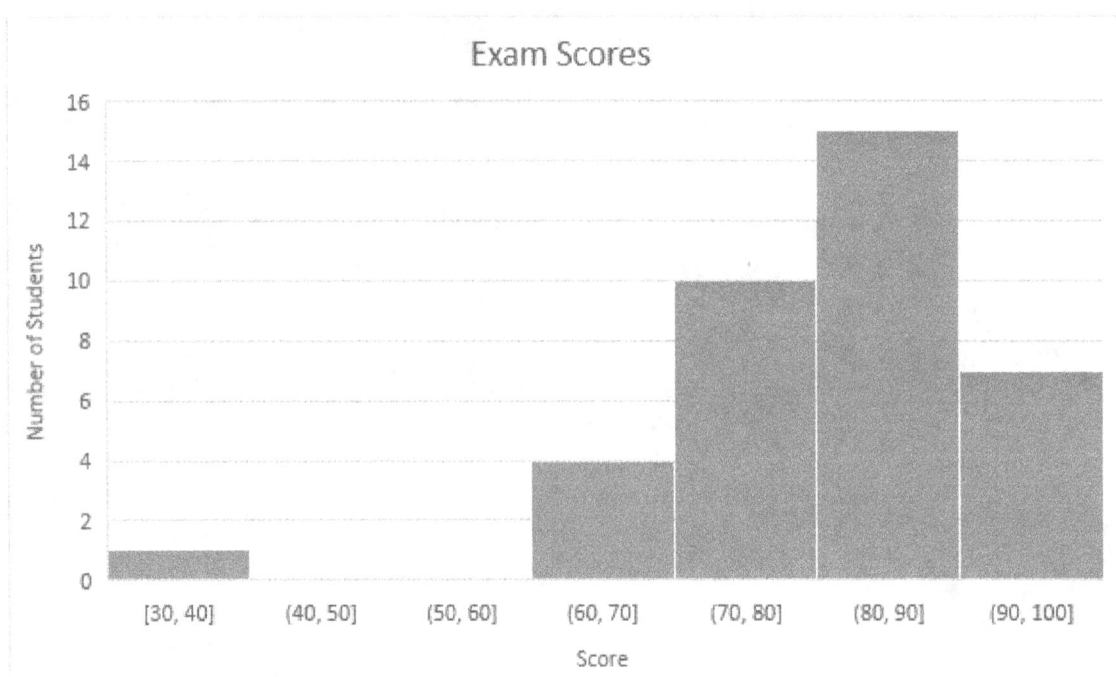

Identify the types of graph(s) that would be appropriate:

1. Tracking the weight of a puppy each month for the first year of his life.

2. Graphing favorite color. Each person is required to vote, and each person may vote only once.

3. Displaying the number of students in each major. Students may be double majors.

4. Showing the heights of students.

5. Showing the rate of growth of a bean plant

6. Four popular books are on a list. Students check off the ones they have read. We wish to display how many students have read each book.

7. Showing the ages of people at a Democratic convention.

Data Shapes

Bell or normal:

Skewed to the left:

Skewed to the right:

Measures of center

Mean:

Median:

Stem and Leaf Plots

Create a stem and leaf plot from the following data sets:

The times, in minutes, for a random sample of runners to complete a race are as follows:

| 36 | 47 | 56 | 41 | 36 | 19 | 48 | 42 | 55 | 23 | 28 | 35 |

| 38 | 37 | 61 | 31 | 43 | 44 |

The shape of the data set is:

The appropriate measure of center is:

The mean will be:

The scores on a recent exam are as follows:

77   79   84   93   92   33   68   78   76   75   52   87

61   81   83   72   85

The shape of the data set is:

The appropriate measure of center is:

The mean will be:

The weights of a random sample of men are recorded as follows:

| 191 | 186 | 163 | 151 | 198 | 172 | 185 | 197 | 204 | 217 | 278 |
| 177 | 222 | 193 | 188 | 181 | 205 | 216 | 168 | 208 | 192 | 179 |

The shape of the data set is:

The appropriate measure of center is:

The mean will be:

Creating Histograms on the Calculators

TI 83/4
Create a list of the values
Hit [2nd] [y=] to access the STATPLOT feature
Select a plot and hit [ENTER]
Turn the plot ON
Using the arrow keys, select the histogram option
Select the appropriate list for Xlist
Press [GRAPH]
Press [ZOOM] [9] to center the histogram

TI Nspire
Type the values into a list and name the list
Hit [ON/Home]
Choose the pink Data & Statistics option along the bottom
Click the X-axis (using the navigation pad) and select the column heading
Hit [MENU]
Select 1: Plot Type
Select 3: Histogram

CASIO
Enter the values into a list
Select [F1] GRPH
Select the graph you want to use and then hit [F6] to SET it up
Arrow down to Graph Type and select [F6] for more options
Select [F1] for Histogram
Select the appropriate list and hit [EXIT]
Select the graph you chose earlier
Pick a starting point (less than or equal to the minimum) and a column width
Select Draw

Creating Box and Whisker Plots on the Calculators

TI 83/4
Create a list of the values
Hit [2nd] [y=] to access the STATPLOT feature
Select a plot and hit [ENTER]
Turn the plot ON
Using the arrow keys, select the box and whisker with outliers option
Select the appropriate list for Xlist
Press [GRAPH]
Press [ZOOM] [9] to center the histogram

TI Nspire
Type the values into a list and name the list
Hit [ON/Home]
Choose the pink Data & Statistics option along the bottom
Click the X-axis (using the navigation pad) and select the column heading
Hit [MENU]
Select 1: Plot Type
Select 2: Box Plot

CASIO
Enter the values into a list
Select [F1] GRPH
Select the graph you want to use and then hit [F6] to SET it up
Arrow down to Graph Type and select [F6] for more options
Select [F2] for Box Plot
Select the appropriate list and hit [EXIT]
Select the graph you chose earlier
Pick a starting point (less than or equal to the minimum) and a column width
Select Draw

Shape	Center	Spread

Mean:

Median:

When computing the mean with an outlier present:

What does an outlier do to the range?

Find the median, 5# summary, range and IQR

Stem	Leaves
1	4
2	
3	
4	6
5	46
6	0273
7	4418
8	2

1. Put the numbers in order

14   46   54   56   60   62   63   67   71   74   74   78   82

2. Find the median

a) $\dfrac{n+1}{2} = \dfrac{13+1}{2} = \dfrac{14}{2} = 7$ 7th slot is the median

14   46   54   56   60   62   **63**   67   71   74   74   78   82

3. If the median is an actual value, cross it out. If the median is "between" two numbers, draw a vertical line between them to separate the halves. Then average the two numbers together to get the value of the median.

14   46   54   56   60   62   **6|3**   67   71   74   74   78   82

4. Look at only the LEFT half of the data set and repeat steps 1 and 2 to find the first quartile.

$$\frac{n+1}{2} = \frac{6+1}{2} = 3.5 \qquad\qquad 3.5^{th} \text{ slot is the first quartile}$$

14   46   54 | 56   60   62

       Since the first quartile is between two values, average those values together to get the value of Q1

$$\frac{54 + 56}{2} = 55$$

5. Repeat the same process on the right side to get the third quartile

$$\frac{n+1}{2} = \frac{6+1}{2} = 3.5 \qquad\qquad 3.5^{th} \text{ slot is the third quartile}$$

67   71   74 | 74   78   82

       Since the first quartile is between two values, average those values together to get the value of Q3

$$\frac{74 + 74}{2} = 74$$

Five number summary:

Min: **14**

Q1:   **55**

Med: **63**

Q3:   **74**

Max: **82**

Range = Max – Min = 82 – 14 = 68

IQR = Q3 – Q1 = 74 – 55 = 19

Median and five number summary example

A 25 year old woman recently went on match.com, and she recorded the ages of the men who contacted her, in the order they were received. The ages were as follows:

25   27   28   24   23   33   30   20   72   23   28   25
25   26   31   27   26   28   29   26

First create a stem and leaf plot to determine the shape of the data (or make a histogram/box and whisker plot on the calculator.)

Since this data is skewed (has an outlier,) we want to find the median as the measure of center.

First step: Rank the data from smallest to largest.

L(M) = the location of the median, and it is found by:

$$L(M) = \frac{n+1}{2}$$

Where n is the sample size. Find the location of this median.

Find the value of the median.

Suppose another value gets added to this data set:

20 23 23 24 25 25 25 26 26 26 27 27 28 28 28 29 30 31 33 **40** 72

Now find the L(M) and the value of the median.

Using the original data set, find the five number summary:

20 23 23 24 25 25 25 26 26 26 27 27 28 28 28 29 30 31 33 72

Range:

IQR:

Finding the five number summary on a graphing calculator

TI 83, 84

Enter data into one of the lists.

Press the [STAT] button again and arrow over to CALC.

Select option 1 and hit [ENTER].

Select which list (L_1 or L_2 for instance) you would like analyzed. L_1 is the default if left blank.

Hit [ENTER]

You will be given the mean, standard deviation and five number summary.

TI Nspire

Enter the data on a List and Spreadsheet page

Name the list

Press [MENU], #4 Statistics, #1 Stat Calculations and choose #1 One-Variable Statistics

"One Variable Statistics" gives the five number summary in the spreadsheet

Casio

Press [MENU] and select STAT

Find out which file you are in by going to SETUP (Shift menu) and scroll down to List File

To change files, press FILE (F1) and the file number, then EXE.EXIT

To ENTER DATA, go to the file and type in the values, hitting EXE after each one

Select CALC, (F2) SET (F6)

The list that holds the values needs to be entered into 1VarXList, press EXE

For 1VarFreq, press F1 to get 1, then press EXIT

Now press 1Var (F1)

The weight losses, in pounds, for a group of 36 dieters were recorded and are as follows:

26	25	19	13	28	20	42	21	21	17	33	23
23	24	26	26	18	38	18	12	27	15	28	19
15	29	17	31	25	12	27	24	16	30	24	19

A) **Create a stem and leaf plot for the data or determine shape on the calculator.**

B) **What shape is the data?**

C) **State the five number summary, range and IQR. Feel free to use your technology.**

Mean example

The IQs of a randomly selected group of people were recorded and are as follows:

| 106 | 100 | 84 | 119 | 98 | 101 | 120 | 79 | 105 | 114 | 99 | 100 |

| 96 | 83 | 132 | 119 | 111 | 91 | 101 | 97 | 113 | 102 | 103 |

Create a stem and leaf plot of the data or a histogram/box and whisker plot on the calculator:

What shape is it?

What measures of center and spread are appropriate?

Determine those values.

For the following examples, create a stem and leaf plot. Then describe the data appropriately (shape, center, spread)

1. For people with a particular disease, the ages at death were recorded for a random sample of 12 people.

    68   78   82   75   21   91   77   69   85   94   76

Shape: **Center:**   **Spread:**

2. **The number of baby sea turtles that made it to water was recorded from 16 randomly selected hatching grounds. They were as follows:**

38	35	46	45	51	41	23	39	43	54	48	32
26	63	37	42								

Shape: **Center:**   **Spread:**

Drawing and labeling normal (bell) curves
68-95-99.7 rule

The students in the day section of a large stat class took an exam, and the scores had the following distribution:

Bell curve   Mean = 70   Standard deviation = 10.

Draw and label that normal curve:

The evening section had the following distribution:

Bell curve   Mean = 70   Standard deviation = 2

Draw and label that normal curve:

The 68-95-99.7 rule:

1. 68% of the data falls within 1 standard deviation of the mean in each direction

2. 95% of the data falls within 2 standard deviations of the mean in each direction.

3. 99.7% of the data falls within 3 standard deviations of the mean in each direction.

68% of the evening section scored between…

What percent of the day class scored between a 50 and a 90?

Actual to percentile

Suppose exam scores are normally distributed with a mean of 70 and a standard deviation of 10. Draw and label that normal curve.

You scored an 80. You wish to find out what percent of your classmates scored lower than you. Use the 68-95-99.7 rule to determine that percentage.

Using the definition of Z-score, determine what your Z-score would be if you scored an 80 on the test. Look up that number on the table to determine the exact percent of people who scored below you.

Suppose you scored a 50. What is your Z-score? What percent of the people scored below you?

Suppose you scored a 90. Use the information provided (not the picture) to determine your Z-score.

Mean = 70   Std dev = 10   You scored a 90.

Suppose you scored an 83. What percent of the people scored below you? What percent scored above you?

Suppose you scored a 52. What is your percentile?

Examples

IQs are normally distributed with a mean of 100 and a standard deviation of 15.

1. Jacquie's IQ is 105. What percent of people have an IQ below her?

2. Peter's IQ is 88. What percent of people have an IQ below him?

3. Xavier's IQ is 118. What percent of people have an IQ above him?

4. Nathaniel has an IQ of 78. What percent of people have an IQ above him?

5. Isabelle's IQ is 81. What percent of people have an IQ below her?

Percentile to Actual

College entrance exams are normally distributed with a mean of 200 and a standard deviation of 10. In order to get into the honors college, you need to score in the 97.73rd percentile. What score do you need on the test?

What is your Z-score?

Using that information, determine what you need to get on the test and create a formula out of it.

Suppose you just wanted to get into the college, and the minimum requirement is that you score at the 30th percentile. What score would you need?

IQs are normally distributed with a mean of 100 and a standard deviation of 15.

1. What IQ is at the 40th percentile?

2. What IQ is at the 25th percentile?

3. To get into the honors program, your IQ must be at or above the 85th percentile. What IQ marks the 85th percentile?

4. Students in the 10th percentile or lower can receive extra help. What IQ marks the 10th percentile?

5. What IQ marks the 52nd percentile?

Z score transformation flow chart

Actual $\longrightarrow$ Z $\longrightarrow$ Percentile

Percentile $\longrightarrow$ Z $\longrightarrow$ Actual

Percentile definition: The % of the data that is BELOW you

Z score (standardized score) definition: Z score tells you how many standard deviations you are from the mean and in which direction.

Z-score transformations on the calculators:

TI 83/4

Actual to percentile

Hit [STAT] and arrow over twice to TESTS
Select option 1: Z-Test
Select Stats for Input
μ_0: Enter the mean
σ: Enter the standard deviation
$\bar{X}$: Enter the actual value
N: Put in the number 1
Select whether it's above or below with < or >
Select Calculate
Z = the Z-score
P = the probability

Percentile to Actual

Hit [2nd] [VARS] for the DISTR function
Select option 3: invNorm(
Enter the proportion BELOW (with a comma or a down arrow after)
Enter the mean (with a comma or down arrow after)
Enter the standard deviation
Enter
The actual value will appear

Z-score transformations on the calculators:

TI Nspire

Actual to percentile

Under a Calculate screen, hit [MENU]
Select option 6: Statistics
Select option 7: Stat Tests
Select option 1: z Test
Select Stats from the pulldown menu and hit [ENTER]

μ_0: Enter the mean

σ: Enter the standard deviation

X: Enter the actual value

N: Put in the number 1

Select whether it's above or below with < or > in the Alternate Hyp pulldown
"z" is the Z-score
"PVal" is the probability

Percentile to Actual

Under the Calculate menu, select [MENU]
Select option 6: Statistics
Select option 5: Distributions
Select option 3: Inverse Normal
For Area, select proportion BELOW
μ: put the mean
σ: **put the standard deviation**
Select OK
The value will appear in the line

Z-score transformations on the calculators:

Casio
Actual to percentile

Enter STAT section through the menu
Select F3 for TEST
Select F1 for Z
Select F1 for 1-S
Select F2 for VAR
Select F2 or F3 for < or >

μ_0: Enter the mean

$\bar{X}$: Enter the actual value

σ: Enter the standard deviation

N: Put in the number 1

Hit Execute

Z = the Z-score
P = the probability

Percentile to Actual

Enter STAT section through the menu
Select F5 for DIST
Select F1 for NORM
Select F3 for InvN
Select F2 for VAR
For TAIL, select F1 for LEFT
For AREA, enter proportion BELOW
σ: Enter the standard deviation
μ: Enter the mean
Press down arrow and F1 for Calculate

Newborn baby weights are normally distributed with a mean of 7.1 pounds and a standard deviation of 1.8 pounds.

1. What percent of newborns weigh more than 8 pounds?

2. What baby weight marks the 30th percentile?

3. A baby born under 5 pounds needs to go into the intensive care unit. What percent of babies need to go into the intensive care unit?

4. What percent of babies are born weighing over 9 pounds?

5. What baby weight is at the 75th percentile?

6. What baby weight is at the 90th percentile?

7. What baby weight marks the 15th percentile?

8. What percent of babies are born under 6.5 pounds?

Averaging Groups Together

Suppose the average weight for all newborn babies is 7 pounds and the standard deviation is 2 pounds. Draw and label the normal curve for the birth weights of all babies:

If each person in the room was tasked with randomly selecting one baby and determining the weight of that baby at birth, 95% of the people would come back tomorrow and report their baby weighed between...

68% would come back and say their baby weighed between...

Is it within the realm of reason that someone could randomly select a baby that weighed 1 pound?

This is true for INDIVIDUAL babies. What happens if each person in the room is tasked with randomly selecting 16 babies and averaging those birth weights together? **Would a person be likely to come back and say their AVERAGE baby weight for the sixteen babies was 1 pound?**

In order for an AVERAGE to come back as 1 pound, that means that all sixteen babies weighed around 1 pound. If that is the case, then the sample must have been gathered at a premature baby unit and not randomly. **If the sample is gathered randomly, then the average for the sixteen babies will be somewhat close to the real population average of 7 pounds.**

The standard deviation for MEANS is $\sigma/\sqrt{n}$ where σ is the standard deviation for individuals, and n is the size of the sample averaged together

Birth weights are normally distributed with a mean of 7 pounds and a standard deviation of 2 pounds. If I asked each of you to randomly select one baby and determine the weight at birth, 95% of you would come back with a weight between...

If I asked each of you to randomly select 16 babies and average their birth weights together, 95% of you would come back with an AVERAGE weight between...

If I asked each of you to randomly select 100 babies and average their birth weights together, 95% of you would come back with an AVERAGE weight between...

Finding Probabilities When Gathering a Sample

IQs are normally distributed with a mean of 100 and a standard deviation of 15.

Draw and label that curve:

Recall:

What percent of the people have an IQ above 103?

What if I asked you each to take a random sample of 25 people and average their IQs together? Remember, with an average, you increase and decrease by $\sigma / \sqrt{n}$ for the standard deviation. Draw and label the normal curve for those AVERAGE IQs.

Now, what percent of those AVERAGES will be above 103?

On the calculator...

On the calculator, use the sample size for n (instead of 1) and type in the given standard deviation for σ.

Example:

Test scores are normally distributed with a mean of 70 and a standard deviation of 12.

If you were to randomly select 36 students from this class and average their test scores together, what percent of those AVERAGES will be above a 73?

If you randomly select 64 students from the class and average their test scores together, what percent of those AVERAGES would be above a 73?

Probabilities

The percent of the curve that falls above or below a certain point is also the probability you will get a value in that range.

If 10% of the students scored above a 90 on a test, then the probability I will randomly select a student who scored above a 90 is 10%.

Examples:

Marathon times are normally distributed with a mean of 270 minutes and a standard deviation of 38 minutes.

What is the probability that a randomly selected runner will finish in less than 200 minutes?

What is the probability that a randomly selected sample of 20 runners will have an **average** time of less than 260 minutes?

What is the probability that 60 randomly selected people will have an **average** time more than 273 minutes?

What is the probability that a randomly selected runner will take more than 300 minutes?

What is the probability that a randomly selected group of 15 runners will have an **average** time less than 264 minutes?

Example:

Heights of men are normally distributed with a mean of 70 inches and a standard deviation of 3 inches.

What is the probability that a randomly selected male will be less than 65 inches tall?

What is the probability that a randomly selected group of 50 men will have an average height of more than 70.5 inches?

What is the probability that a randomly selected group of 37 men will have an average height less than 69 inches tall?

What height is at the 60th percentile?

Don't forget... (just use 1 for n)

What percent of men are less than 68 inches tall?

What percent of men are more than 65 inches tall?

What percent of men are more than 74 inches tall?

Mutually Exclusive and Independent

Mutually exclusive:

Consider the following:

A class has the following make-up:

Freshmen: 25%
Sophomores: 30%
Juniors: 35%
Seniors:

If the class only consists of undergraduates, what percent are seniors?

If you select one student at random, what is the probability that the student is a freshman or a sophomore?

If you select one student at random, what is the probability that the student is a sophomore or a senior?

If you select one student at random, what is the probability that the student is a freshman and a senior?

If you were informed that the class also contained grad students, what percent of the class are seniors?

Independent:

Consider the following for a class of undergraduates:

Freshmen: 25%	Blue eyes: 30%
Sophomores: 30%	Brown eyes: 50%
Juniors: 35%	Hazel: 15%
Senior:	Other:

Fill in the missing amounts. What enables you to answer that question?

If you pick a student at random, what is the probability that the student will have both brown eyes and sophomore status?

If you pick a student at random, what is the probability that the student will have blue eyes and senior status?

If you pick two students at random, what is the probability that they will both be juniors?

If I tell you a chosen student is a senior, do you know that student's eye color?

	Key word	Operation
Mutually exclusive		
Independent		

Examples:

In a recent survey, the following information was gathered among employees at a local company:

College educated: 49%

Own pets: 76%

Married: 62%

Has no children: 12%

Has one child: 24%

Has two children: 56%

If I select one employee at random, what is the probability that employee is married and has one child?

If I select one employee at random, what is the probability that the employee has one or two children?

If I select one employee at random, what is the probability that the employee has no pets?

If I select two employees at random, what is the probability that they are both college educated (assume replacement?)

If I select an employee at random, what is the probability that they have three or more kids?

Example 2

A bag of candy contains five flavors. The percent of the bag that is each flavor is as follows:

Lemon: 10%

Strawberry: 20%

Grape: 30%

Lime: 10%

Cherry:

What percent must be cherry?

If you select one candy at random, what is the probability it is grape or lime?

If you select one candy at random, what is the probability it is not strawberry?

If you select two candies at random, with replacement, what is the probability they are both grape?

If you select one candy at random, what is the probability you will select lemon or strawberry?

If you select two candies at random, with replacement, what is the probability they will both be cherry?

Risk and Relative Risk:

Suppose a construction company has improved its safety guidelines. Workers are required to wear a helmet that can withstand a more substantial impact. With the same type of accident, the risk of getting injured with the old helmet is 30%, whereas the risk of getting injured with the new helmet is just 10%?

What is the relative risk of getting injured with the old helmet compared to the new?

Relative risk definition:

The risk of getting lung cancer for smokers is 48%. The risk for getting lung cancer for non-smokers is 2%. What is the relative risk for getting lung cancer for smokers relative to non-smokers? What does that mean?

The risk of having a bad reaction to an old vaccine is 1.2%. The risk of having a bad reaction to the new vaccine is 0.4%. What is the relative risk of having a reaction on the old vaccine compared to the new? What does that mean?

Simulation

Simulation involves using random outcomes to simulate chance behavior:

Simulation can be used to represent something with two equally likely outcomes, like the gender of a baby. You can use coin flipping to simulate the birth of children.

Assign "heads" to mean girl, and "tails" to mean boy. Simulate the birth of 5 children using an online coin flipper. What genders did you have, and in what order?

Suppose half of the students in your dorm are majoring in the humanities and sciences. Use coin flipping to simulate the random selection of 10 students from your dorm, making a note of how many are majoring in the humanities and sciences.

Suppose the odds of failing this class are 1 out of 6. Use dice to simulate the random selection of 8 students, determining how many of your sample will fail?

Suppose a basketball player makes 83.333% of his free throws. (That equates to 5/6.) Simulate 10 free throws, stating how many the player makes.

20% of people have cavities. If a dentist sees 18 people in a day, use a random number generator to simulate 18 patients, noting how many of them had cavities.

3 out of 10 people will be in a car accident in their lifetime. Use a random number generator to simulate the random selection of 12 people, determining how many have been in (or will be) in a car accident.

When to use a coin:

When to use dice:

When to use a random number generator:

Confidence Intervals on Proportions

A random sample of 100 men revealed that 20% would seek treatment if they encountered hair loss. Find a 95% confidence interval for the proportion of men who would seek treatment for hair loss.

1. Find the sample proportion p.

2. If the assumptions are met, compute the standard deviation by using the formula. Once you have a p value and a standard deviation value, you can complete a normal curve. (You know it's a normal curve because of the test in step 2.) Put the p value in the middle of the curve and increase/decrease by the standard deviation value.

You can be roughly 95% confident that between _____ and _____ percent of all men would seek treatment for hair loss.

What about 68% confident?

Make a confidence statement, which has the following three parts:

a) state how confident you are
b) tie it to the problem, making sure to reference "all" or "population"
c) state the interval

How to do confidence intervals on proportions on the calculators

TI 83/84
Press [STAT]
Arrow over twice to TESTS
Select option A, 1-PropZInt...
Type in HOW MANY said yes for x
Type in the sample size for n
Determine the level of confidence (as a decimal)
Calculate

TI Nspire
Press [MENU]
Select option 6: Statistics
Select option 6: Confidence Intervals
Select option 5: 1-Prop Z Interval
Successes x: is HOW MANY said yes
n = sample size
C level is the confidence level as a decimal
Press [OK]

Casio
From the list menu, select F4, INTR
Select F1 for Z
Select F3 for 1-P
Type in the confidence level as a decimal
For x, type in HOW MANY said yes
For n, type in the sample size
Execute

1. A recent study of 200 randomly selected pregnant women showed that 140 had side effects from the iron supplements. Find a 99% confidence interval for the proportion of all pregnant women who suffer side effects from iron supplements. Write the confidence statement.

2. A random sample of 1000 shoppers showed that 880 are more likely to shop if there is a store-wide sale. Calculate and interpret a 99% confidence interval for the proportion of all shoppers who are more likely to shop during a store-wide sale.

3. A random sample of 200 people revealed that 62% are unable to drive a stick shift. Calculate and interpret a 90% confidence interval for the proportion of all people who are unable to drive a stick shift.

How to covert a percentage into "how many:"

Always multiply the percentage (as a decimal) by the sample size. Remember, it only needs to be done when you are given a percentage.

4. A random sample of 300 veterinarians showed that 52 of them recommended "Brand A" flea control. Calculate and interpret a 95% confidence interval for the proportion of all veterinarians who recommend "Brand A."

5. A random sample of 500 teachers showed that 76% voted Democratic in the last election. Calculate and interpret a 99% confidence interval for the proportion of all teachers who voted Democratic in the last election.

What does a confidence interval mean?

Coin activity:

Use a simulator to flip your coin 20 times and determine what proportion of the time "heads" appeared.

What is the value of your sample proportion $\hat{p}$?

Can you say conclusively that this has to be the population proportion p?

Calculate a 95% confidence interval for the true proportion (p) of all coin flips that will land heads up.

Make a confidence statement.

What percent of your calculated intervals should contain the true proportion of 0.5? What percent should miss?

M&M Activity

Of interest is to determine what proportion of all M&Ms are orange. Once you receive your package of M&Ms, pull out the first 50, counting how many of them are orange. Use that information to make a 90% confidence interval for the proportion of all M&Ms that come out of the factory orange.

Number of orange M&Ms	Lower Bound	Upper Bound	Number of orange M&Ms	Lower Bound	Upper Bound
1	-0.0126	0.0526	30	0.4860	0.7140
2	-0.0056	0.0856	31	0.5071	0.7329
3	0.0048	0.1152	32	0.5283	0.7517
4	0.0169	0.1431	33	0.5498	0.7702
5	0.0302	0.1698	34	0.5715	0.7885
6	0.0444	0.1956	35	0.5934	0.8066
7	0.0593	0.2207	36	0.6156	0.8244
8	0.0747	0.2453	37	0.6380	0.8420
9	0.0906	0.2694	38	0.6607	0.8594
10	0.1070	0.2931	39	0.6836	0.8764
11	0.1236	0.3164	40	0.7070	0.8931
12	0.1407	0.3394	41	0.7306	0.9094
13	0.1580	0.3620	42	0.7547	0.9253
14	0.1756	0.3844	43	0.7793	0.9407
15	0.1934	0.4066	44	0.8044	0.9556
16	0.2115	0.4285	45	0.8302	0.9698
17	0.2298	0.4502	46	0.8569	0.9831
18	0.2483	0.4717	47	0.8848	0.9952
19	0.2671	0.4929	48	0.9144	1.0060
20	0.2860	0.5140			
21	0.3052	0.5348			
22	0.3245	0.5555			
23	0.3441	0.5759			
24	0.3638	0.5962			
25	0.3837	0.6163			
26	0.4038	0.6362			
27	0.4241	0.6559			
28	0.4445	0.6755			
29	0.4652	0.6948			

Confidence Intervals on Means

A random sample of 25 VCU students had an average height of 68 inches. The standard deviation for heights is known to be 4 inches. Calculate and interpret a 95% confidence interval for the AVERAGE height of all VCU students.

Use the procedure on the next page to determine the interval:

Create the confidence statement:

Does this mean that ALL VCU students are in that height range?

Does this mean that 95% of VCU students are in that height range?

It means that if you measured the height of EVERY VCU student and AVERAGED their heights together, you believe the AVERAGE would be in that range.

How to do confidence intervals on means on the calculators

TI 83/84
Press [STAT]
Arrow over twice to TESTS
Select option 7, ZInterval...
Type in the standard deviation for σ
Type in the sample average for $\bar{x}$
Type in the sample size for n
Determine the level of confidence (as a decimal)
Calculate

TI Nspire
Press [MENU]
Select option 6: Statistics
Select option 6: Confidence Intervals
Select option 1: z Interval
Select Stats as Data Input Method
$\bar{x}$ = sample mean
σ_x = standard deviation
n = sample size
C level is the confidence level as a decimal
Press [OK]

Casio
From the list menu, select F4, INTR
Select F1 for Z
Select F1 for 1-s
For Data: Select Variable
Type in the confidence level as a decimal
For $\bar{x}$ type in the sample mean
For σ_x type in the standard deviation
For n, type in the sample size
Execute

A random sample of 200 VCU students had an average SAT score of 1580. The standard deviation for SATs is 300. Calculate and interpret a 90% confidence interval for the average SAT score for all VCU students.

A random sample of 65 writers showed their first published manuscript had been rejected an average of 35.6 times with a standard deviation of 4.2. Calculate and interpret a 99% confidence interval for the average number of rejections for new authors.

A random sample of 700 college students revealed that they spent an average of $585.76 on books in one semester. The standard deviation was $135.09. Calculate and interpret a 90% confidence interval for the average amount spent on books per semester for all college students.

Confidence Intervals on means given data

A random sample of 15 dieters had the following weight losses (in pounds) after 6 months on a diet plan.

15.2   14.9   11.5   19.2   9.3   16.9   11.0   18.4   8.2   13.4   11.7   13.4   15.6

10.2   18.3

Determine the mean and standard deviation from the sample.

Use the T interval to calculate and interpret a 90% confidence interval for the mean weight loss for all people on this plan.

Circle all of the sample weight losses that fell OUTSIDE of that range. Does this interval mean that 90% of weight losses will be in that range?

A random sample of 22 third graders at Gleason Elementary had their SOL scores recorded. They were as follows:

| 425 | 460 | 480 | 526 | 571 | 540 | 379 | 411 | 502 | 403 | 460 |
| 495 | 433 | 391 | 600 | 546 | 439 | 477 | 521 | 540 | 488 | 469 |

Calculate and interpret a 95% confidence interval for the average SOL score for all third graders at Gleason Elementary.

Dice activity

Use software to roll a die 25 times, recording your roll each time. Put the rolls in this table.

Using that data, calculate and interpret a 95% confidence interval for the mean dice roll for all dice rolls.

Determining the difference in confidence interval types:

The key words to look for are **PROPORTION/PERCENT** vs. **AVERAGE/MEAN.**

Use the correct procedure to determine the following confidence intervals:

1. A random sample of 600 new mothers showed that 82% are breast-feeding. Calculate and interpret a 95% confidence interval for the proportion of all new mothers who are breastfeeding.

2. A random sample of 400 young adults revealed that they had been to the emergency room an average of 3.2 times between birth and age 18. The standard deviation was 1.3. Calculate and interpret a 99% confidence interval for the average number of emergency room trips from birth to age 18.

3. A random sample of 10 batteries had the following life spans in minutes.

482   503   499   479   504   510   495   488   500   496

Calculate and interpret a 95% confidence interval for the average life span of all batteries.

4. In order to get an idea of what proportion of people take a daily multivitamin, a random sample of 300 people was gathered. Of those, 164 took a daily multivitamin. Calculate and interpret a 95% confidence interval for the proportion of all people who take a multivitamin.

5. A random sample of 300 VCU students showed that 72 had withdrawn from a class. Calculate and interpret a 90% confidence interval for the proportion of all VCU students who have withdrawn from a class.

6. A random sample of 400 people on "Medication A" showed an average blood pressure increase of 15.2 points, with a standard deviation of 3.7 points. Calculate and interpret a 99% confidence interval for the average blood pressure increase for people on Medication A.

7. A random sample of 200 dental patients showed that 13% needed to come back for additional services. Calculate and interpret a 95% confidence interval for the proportion of all dental patients who need to come back for additional services.

8. A random sample of 11 textbooks was selected from the bookstore, and their prices (in dollars) were as follows:

152   202   245   301   190   229   267   198   275   224   189

(Doesn't this make you love this book?)

Calculate and interpret a 90% confidence interval for the average cost for all textbooks in that store.

Hypothesis Testing on Proportions

Step 1. Write the null and alternative hypotheses

Null: Always contains an = sign and CONTAINS THE VALUE FOR "p"
States the current belief in the population
Alternative: Always contains an <,>, or ≠
States what the researcher believes will be true.

Step 2. Determine p-hat BASED ON THE STUDY

P-hat will be given as a decimal, percent, or "out of" statement

Step 3. Plug p and p-hat into the formula:

$$Z = \frac{\hat{p} - p}{\sqrt{\dfrac{p(1-p)}{n}}}$$

Always use the value from the null to determine the standard deviation in the denominator.

***THIS FORMULA FOLLOWS THE FAMILIAR FORM

$$Z = \frac{\underline{OBSERVED\ SCORE - MEAN}}{STD.\ DEV}$$

Step 4. Look up the Z-score on the chart to find the associated percentile

Step 5. Find the P-value

 If the alternative hypothesis in step 1 is a "less than" statement, the percentile is the p-value.

 If the alternative hypothesis in step 1 is a "greater than" statement, you must subtract the percentile from 1 to get the p-value

 If the alternative hypothesis in step 1 is a "not equal to" statement, then look up the negative version of the Z-score and double the percentile to get the p-value.

Step 6. Draw the conclusion

If the p-value is less than the significance level given in the problem, then reject the null in favor of the alternative. Acknowledge you may have made a type I error.

If the p-value is greater than the significance level given in the problem, then fail to reject the null. Acknowledge you may have made a type II error.

In class example for hypothesis testing on proportions

A study done in 1980 revealed that 50% of teens aged 16-19 were sexually active. You believe the current percentage is higher than that. You plan to randomly select 100 teens in that age bracket and determine if they are sexually active.

You must begin under the assumption that the previous finding is still true unless you can prove otherwise. If that's the case, 50% of the population would still be sexually active today (notice that value would be p and not p-hat). You should put that value in the center of a normal curve. Find the value of the standard deviation.

Draw the appropriate normal curve.

If you conduct your survey and come up with a percentage that is in line with the values in the above normal curve, then your data supports the claim from 1980. If the percentage from your sample is not aligned with the values in the normal curve above, then your study is contradicting the study from 1980, and you have what is called a *significant finding*.

A study done in 1980 revealed that 50% of teens aged 16-19 were sexually active. You believe the current percentage is higher than that. You survey 100 randomly selected teens and discover that 61% are sexually active. Are these findings significant at the .05 level?

Null Hypothesis:

Alternative Hypothesis:

P-hat value:

Test statistic: This tells you how many standard deviations you are from the mean of the normal curve on the previous page.

P-value: This tells you how likely you would be to get your sample value if the 1980 information were still true. (What % chance do you have of getting a sample that is 61% sexually active if the actual population percentage is 50% sexually active?)

Conclusion:

Hypothesis Testing on Proportions Using Calculators:

Using the same example, here are the steps to do a hypothesis test with technology:

Step 1: Write the null and alternative hypotheses

Null: p = .5

Alternative: p > .5

Step 2: Determine the z score and the p-value on the calculator (next page)

Null value = Po = .5

Successes = x = .61 * 100 = 61 people

n = sample size = 100

Direction of alternative is >

Z = 2.2

P-value = .0139

Step 3: Plot the p-value on the world's greatest picture

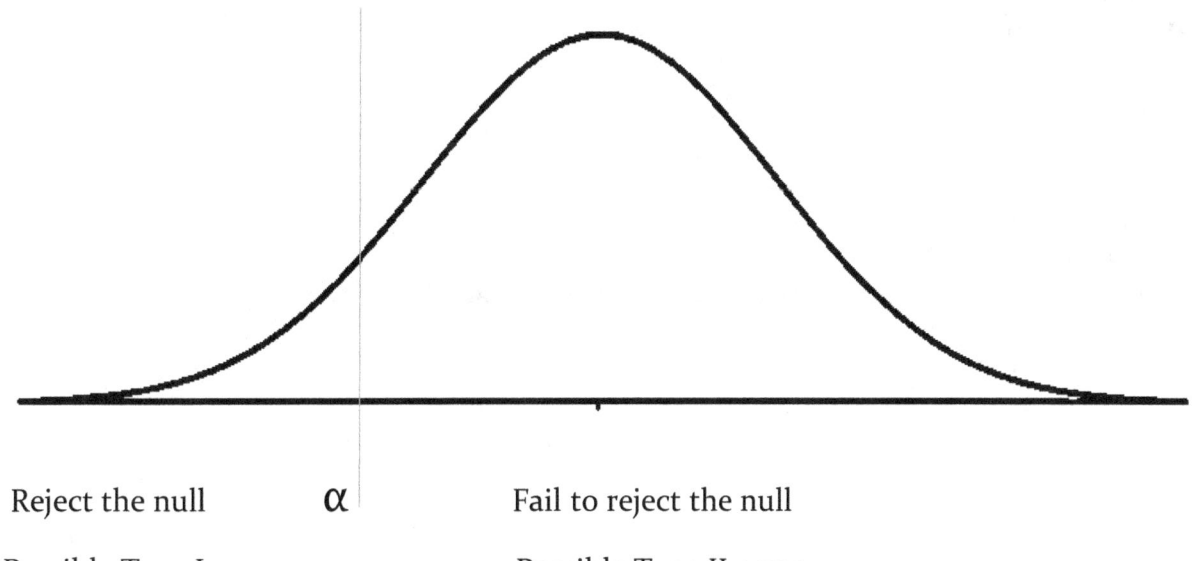

Reject the null   α   Fail to reject the null

Possible Type I error   Possible Type II error

115

Hypothesis Testing on Proportions on the Calculators

TI 83/4
Press [STAT]
Arrow over twice to TESTS
Select option 5: 1-PropZTest
Po = value from null
x = number of successes
n = sample size
Select whether the alternative hypothesis had $\neq$, < or >
Select Calculate
z = Z score
p = p-value

TI Nspire
Press [menu]
Select option 6: Statistics
Select option 7: Stat Tests
Select option 5: 1-Prop z Test
Po = value from null
Successes, x = number of successes
n = sample size
Select whether the alternative hypothesis had $\neq$, < or >
Select OK
"z" = Z-score
"PVal" = p-value

Casio
From the STAT menu, select F3 for Test
Select F1 for Z-test
Select F3 for 1 – Prop
Prop: Select whether the alternative hypothesis had $\neq$, < or >
po = value from null
x = number of successes
n = sample size
Select Execute
z = Z score
p = p-value

An existing drug has 85% effectiveness; you have come up with a new drug that you think is more effective. In a study you conducted, you found 182 out of 200 randomly selected people showed improvement on your new medication. Is this significant at the .05 level?

1. State the hypotheses:

Null:

Alternative:

    2. Determine the z-score

    3. State the p-value

    4. Plot the p-value on the world's greatest picture

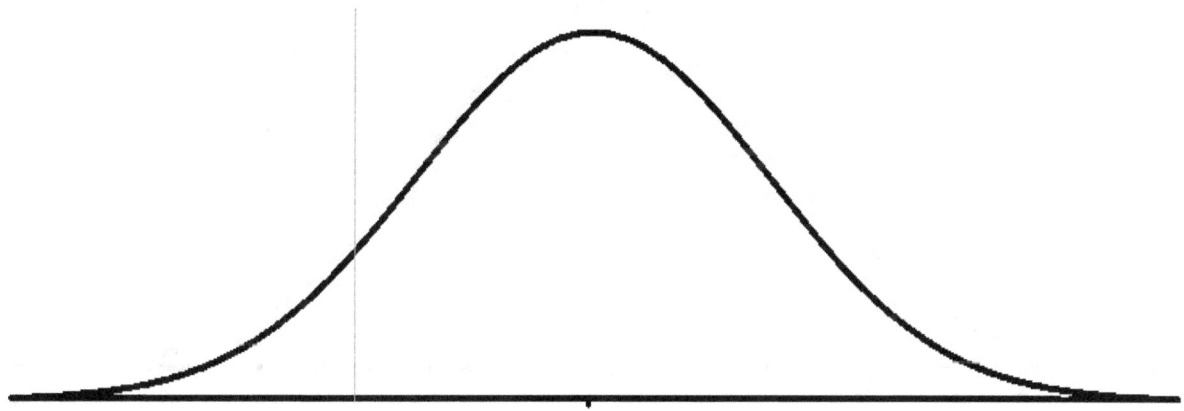

    5. State your conclusion

    6. State possible errors, if asked

A researcher wants to show that fewer than 23% of teens smoke. In a random sample of 100 teens, 18% claim they are smokers. Test the hypothesis at the .01 level.

1. State the hypotheses:

Null:

Alternative:

2. Determine the z-score

3. State the p-value

4. Plot the p-value on the world's greatest picture

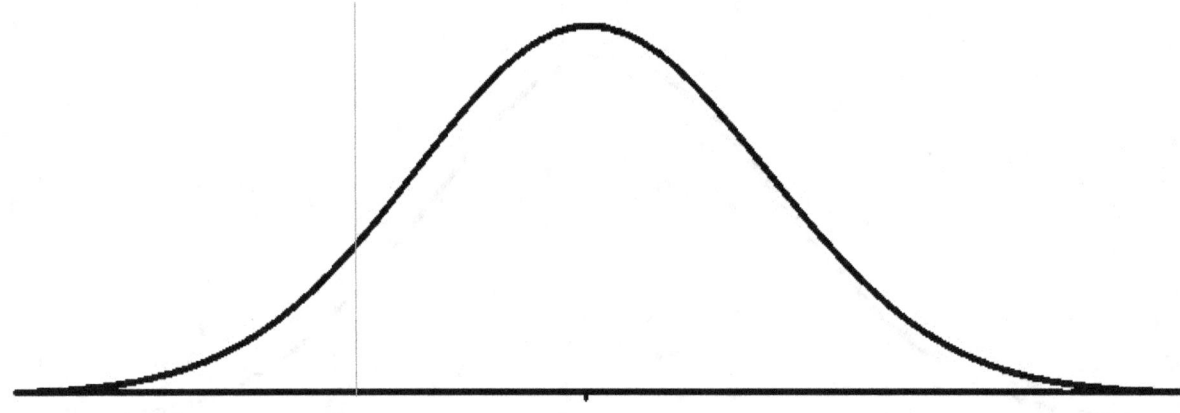

5. State your conclusion

6. State possible errors

An old study showed that 18% of teens have tried smoking. You believe it is different than that. A random sample of 200 teens showed that 32 of them have tried smoking. Is this significant at the .02 level?

You believe that more than 80% of people will feel better on your new drug. A random sample of 400 people showed that 82% felt relief. Is this significant at the .05 level?

Hypothesis testing on means with sigma known

Step 1. Write the null and alternative hypotheses

    Null: Always contains an = sign and CONTAINS THE VALUE FOR "μ "
    Alternative: Always contains an <,>, or $\neq$

Step 2. Determine $\overline{X}$ and s BASED ON THE STUDY and check the assumptions.

Step 3. Plug $\overline{X}$, μ, σ and n into the formula

$$z = \frac{\overline{X} - \mu}{\sigma / \sqrt{n}}$$

Step 4. Look up the Z-score on the chart to find the associated percentile

Step 5. Find the P-value

    If the alternative hypothesis in step 1 is a "less than" statement, the percentile is the p-value.

    If the alternative hypothesis in step 1 is a "greater than" statement, you must subtract the percentile from 1 to get the p-value

    If the alternative hypothesis in step 1 is a "not equal to" statement, then look up the negative version of the Z-score and double the percentile to get the p-value.

Step 6. Draw the conclusion

If the p-value is less than the significance level given in the problem, then reject the null in favor of the alternative. Acknowledge you may have made a type I error.

If the p-value is greater than the significance level given in the problem, then fail to reject the null. Acknowledge you may have made a type II error.

The heights of women in this country are normally distributed with a mean of 64 inches and a known standard deviation of 2.5 inches. You want to see if a group of women with "Medical condition A," on average, shorter than American women as a whole. You randomly select 17 women with this condition, and their average height is 62.7 inches. Is this finding statistically significant at the .01 level?

Hypotheses:

Test Statistic:

P-Value:

Decision:

Hypothesis Testing on Means on the Calculators

TI 83/4

Press [STAT]

Arrow over twice to TESTS

Select option 1: Z-Test

Select Stats

μ_0 = value from null

σ = standard deviation

$\bar{x}$ = sample mean

n = sample size

Select whether the alternative hypothesis had $\neq$, < or >

Select Calculate

z = Z score

p = p-value

TI Nspire

Press [menu]

Select option 6: Statistics

Select option 7: Stat Tests

Select option 1: Z-Test

Select Stats as the input method

μ_0 = value from null

σ = standard deviation

$\bar{x}$ = sample mean

n = sample size

Select whether the alternative hypothesis had $\neq$, < or >

Select OK

"z" = Z-score

"PVal" = p-value

Hypothesis Testing on Means on the Calculators

Casio
From the STAT menu, select F3 for Test
Select F1 for Z-test
Select F1 for 1-sample
Prop: Select whether the alternative hypothesis had $\neq$, < or >
μ_0 **= value from null**
$\bar{x}$ **= sample mean**
σ = standard deviation
n = sample size
Select Execute
z = Z score
p = p-value

The average age of a woman at the birth of her first child 27, with a known standard deviation of 3. You believe that women with alcoholic fathers have a higher average than 27. You randomly select 100 women with alcoholic fathers and determine their average age at the birth of their first child is 27.8. Is this significant at the .05 level?

Hypotheses:

Test Statistic:

P-Value:

Decision:

You believe the average SAT score at your school is above 1500. A random sample of 200 students had an average score of 1524.6 with a standard deviation of 164.7. Is this significant at the .01 level?

Hypotheses:

Test Statistic:

P-Value:

Decision:

Hypothesis testing on means given data

A recent study claimed that the average SOL score for students at a particular school is 475. You believe it is higher than that. A random sample of 15 students from the school revealed the following SOL scores:

405   460   510   570   422   430   390   600   580   550   480   510   530
525   490

Test the hypothesis at the .05 significance level.

Hypotheses:

Test Statistic:

P-value:

Decision:

A politician claimed the average tax cut on his plan would be $500, but you believe it is actually lower than that. A random sample of 18 people revealed the following tax cuts:

20   75   110   160   180   210   335   340   380   424   438   465

470   500   511   523   549   1096

Test this hypothesis at the .05 significance level.

Hypotheses:

Test Statistic:

P-value:

Decision:

Mixed examples:

1. A recent article declared that 56% of Americans approve of the way the president is handling a particular crisis. You believe it is less than that. A random sample of 1500 Americans revealed that 795 approved. Is this significant at the .05 level?

2. The same article declared that the average tax break for Americans was $252. You think the average tax break was lower than that. (You don't like the president.) A random sample of 100 Americans revealed their average tax break was $239 with a standard deviation of $82. Is this significant at the .05 level?

3. A company claimed their average wait time for customer service was 2 minutes. You believe the average is higher than that. A random sample of 20 customers revealed the following wait times. Test the hypothesis at the .01 level.

2	3	3	5	1	2	0	3	6	0
8	1	4	3	2	0	0	3	5	1

4. The average IQ of all people is 100 with a standard deviation of 15. You believe the average IQ at VCU is higher than that. A random sample of 300 VCU students revealed an average IQ of 102.5. Is this significant at the .01 level?

5. An article claimed that, nationally, 18% of college students drop out before graduation. You think the value at VCU is lower than that. A random sample of 200 people who came to VCU as freshmen six years ago was collected, and 16% of them had dropped out. Is this significant at the .02 level?

6. An article claimed the average rent for a 1-bedroom apartment is $820, but you believe it is higher than that. A random sample of 10 apartments revealed the following rent charges:

750   860   975   695   780   820   845   900   910   860

Is this significant at the .05 level?

Hypothesis Testing Activity #1:

The common belief is that people do not have ESP, and, therefore, they cannot read minds. You wish to test to see if you can read my mind. The way you will do this is to tell me the suit of the card I am looking at when I randomly pull a card from a standard deck. We will repeat this 24 times, and each time you will predict the suit I am looking at. Use a significance level of .05 for this test.

Hypotheses:

Sample collection:

Card #	1	2	3	4	5	6	7	8	9	10	11	12
Spades												
Hearts												
Clubs												
Diamonds												
Correct?												

Card #	13	14	15	16	17	18	19	20	21	22	23	24
Spades												
Hearts												
Clubs												
Diamonds												
Correct?												

Z score:

P-value:

# correct	Z score	p-value	
1	-2.357	0.9908	
2	-1.8856	0.9703	
3	-1.4142	0.9214	
4	-0.9428	0.8271	
5	-0.4714	0.6813	
6	0	0.5	
7	0.4714	0.3817	
8	0.9428	0.1729	
9	1.1412	0.0786	
10	1.8856	0.0297	3 out of 100
11	2.357	0.0092	9 out of 1,000
12	2.8284	0.0023	2 out of 1,000
13	3.2998	0.0005	5 out of 10,000
14	3.7712	8 out of 100,000	
15	4.2426	1 out of 100,000	
16	4.714	1 out of a million	
17	5.1854	1 out of ten million	
18	5.6569	8 out of 1 billion	
19	6.1283	4 out of 10 billion	
20	6.5997	2 out of 100 billion	
21	7.0711	7 out of 10 trillion	
22	7.5425	2 out of 100 trillion	
23	8.0139	6 out of 10 quadrillion	
24	8.4853	1 out of 100 quadrillion	

What conclusion do you draw?

What type of error could that be?

Hypothesis Testing Activity #2:

Your stat teacher claims the average roll for a single die is 3.5, and you wish to see if it is actually different than that. Gather a sample by rolling a die 20 times and putting the values in the chart:

Did you prove your point at the .05 significance level?

Hypotheses:

Test statistic:

P-value:

Conclusion:

Examples

Proportions

Find the sample proportions in each instance. Round to four places if appropriate.

1. In a random sample of 500 people, 117 claim they have done an outdoor adventure course.

2. A random sample of 600 people revealed that the proportion of people who think the police should wear body cameras is .8033

3. In a recent random sample of 65 people, 27.69% said they plan to travel for the holidays.

4. A random sample of dog owners showed that 39 out of 56 of them use topical medicine for flea and tick control.

5. 27.8% of a random sample of 1000 drivers knew how to use a stick shift.

6. In a random sample of 645 people, the proportion who suffer from seasonal allergies is .7581.

1.   In a random sample of 500 people, 117 claim they have done an outdoor adventure course.

p-hat = 117/500 = .234   **(out of form)**

2.   A random sample of 600 people revealed that the proportion of people who think the police should wear body cameras is .8033

p-hat = .8033   **(decimal form)**

3.   In a recent random sample of 65 people, 27.69% said they plan to travel for the holidays.

p-hat = .2769   **(percent form)**

4.   A random sample of dog owners showed that 39 out of 56 of them use topical medicine for flea and tick control.

p-hat = 39/56 = .6964   **(out of form)**

5.   27.8% of a random sample of 1000 drivers knew how to use a stick shift.

p-hat = .278   **(percent form)**

6.   In a random sample of 645 people, the proportion who suffer from seasonal allergies is .7581.

p-hat = .7581   **(decimal form)**

Experiments vs. Observational Studies

Identify each of the following as an experiment or an observational study. Indicate if you can declare cause and effect.

1. In order to test the effectiveness of different heartworm pills, a random sample of 100 dogs was randomly divided into four groups. Twenty-five of the dogs each received a different type of heartworm medication. After six months, incidents of heartworms were recorded and compared.

2. Studies have shown that women who drink alcohol while pregnant give birth to babies with developmental delays.

3. A group of students from Milldale High School was randomly selected to participate in an SAT preparation course. After the SATs, the average score of the people who participated in the course was compared to the average score of those who didn't.

4. A group of students who already took an SAT prep course at Milldale High School had their average SAT score compared to those who didn't.

5. A random sample of 1000 dental patients revealed that people who floss had a lower incidence of gum disease than those who didn't.

1.	In order to test the effectiveness of different heartworm pills, a random sample of 100 dogs was randomly divided into four groups. Twenty-five of the dogs each received a different type of heartworm medication. After six months, incidents of heartworms were recorded and compared.

Experiment: the groups were randomly assigned and a treatment was administered. Yes, you can declare cause and effect.

2.	Studies have shown that women who drink alcohol while pregnant give birth to babies with developmental delays.

This had to have been an observational study. Pregnant women would not be randomly assigned to drink during pregnancy. You may not declare cause and effect.

3.	A group of students from Milldale High School was randomly selected to participate in an SAT preparation course. After the SATs, the average score of the people who participated in the course was compared to the average score of those who didn't.

Experiment: the group was randomly chosen and they were told to take the program. Yes, you can declare cause and effect.

4.	A group of students who already took an SAT prep course at Milldale High School had their average SAT score compared to those who didn't.

Observational study: the students came having already taken the course. No behavior change was administered. You may not declare cause and effect, because the students who signed up for the course voluntarily may be the students who struggle most in school, making it an unfair comparison.

5.	A random sample of 1000 dental patients revealed that people who floss had a lower incidence of gum disease than those who didn't.

Observational study: The implication is that the people flossed on their own, without being told. You may not declare cause and effect because the people who floss may also brush more carefully/frequently or they may also use mouthwash.

Vocabulary (1)

State the explanatory and response variables in each instance.

1.   In order to test the effectiveness of different heartworm pills, a random sample of 100 dogs was randomly divided into four groups. Twenty-five of the dogs each received a different type of heartworm medication. After six months, incidents of heartworms were recorded and compared.

2.   Studies have shown that women who drink alcohol while pregnant give birth to babies with developmental delays.

3.   A group of students from Milldale High School was randomly selected to participate in an SAT preparation course. After the SATs, the average score of the people who participated in the course was compared to the average score of those who didn't.

4.   A group of students who already took an SAT prep course at Milldale High School had their average SAT score compared to those who didn't.

5.   A random sample of 1000 dental patients revealed that people who floss had a lower incidence of gum disease than those who didn't.

The explanatory variable is what makes the groups you are comparing different from each other. The response variable is what gets measured and compared.

1. In order to test the effectiveness of different heartworm pills, a random sample of 100 dogs was randomly divided into four groups. Twenty-five of the dogs each received a different type of heartworm medication. After six months, incidents of heartworms were recorded and compared.

Explanatory: Heartworm Pill   **Response: Incidence of Heartworms**

2. Studies have shown that women who drink alcohol while pregnant give birth to babies with developmental delays.

Explanatory: Alcohol consumption   **Response: Developmental delays**

3. A group of students from Milldale High School was randomly selected to participate in an SAT preparation course. After the SATs, the average score of the people who participated in the course was compared to the average score of those who didn't.

Explanatory: SAT prep course   **Response: SAT score**

4. A group of students who already took an SAT prep course at Milldale High School had their average SAT score compared to those who didn't.

Explanatory: SAT prep course   **Response: SAT score**

5. A random sample of 1000 dental patients revealed that people who floss had a lower incidence of gum disease than those who didn't.

Explanatory: Flossing   **Response: Gum disease**

6. A researcher wants to see if premature babies thrive better if the temperature is warmer. The researcher randomly selected ten hospitals in Virginia to use in her study. From those ten hospitals, five were randomly selected to increase the temperatures in the Newborn Intensive Care Units, where the premature babies go after birth. The growth rates of the babies at those five hospitals were compared to the growth rates of the babies at the other five hospitals.

What is the population of interest?

What is the sampling frame?

Who comprises the sample?

Who are the individuals?

What are the variables of interest?

What is the explanatory variable?

What is the response variable?

Is this an experiment or an observational study?

6.   A researcher wants to see if premature babies thrive better if the temperature is warmer. The researcher randomly selected ten hospitals in Virginia to use in her study. From those ten hospitals, five were randomly selected to increase the temperatures in the Newborn Intensive Care Units, where the premature babies go after birth. The growth rates of the babies at those five hospitals were compared to the growth rates of the babies at the other five hospitals.

What is the population of interest?

Premature babies

What is the sampling frame?

Premature babies in NICU units in Virginia

Who comprises the sample?

Premature babies in the selected hospitals

Who are the individuals?

Each premature baby

What are the variables of interest?

Temperature and growth rate

What is the explanatory variable?

Temperature

What is the response variable?

Growth rate

Is this an experiment or an observational study?

Experiment

7. A researcher wants to see the relationship between sleep and performance among VCU students. An email was sent out to all current VCU students, asking them if they'd like to participate in the study. For those who replied, a survey was sent out, asking them to record the number of hours they slept over the course of two weeks. In addition, the students were to report their GPA.

What is the population of interest?

What is the sampling frame?

Who comprises the sample?

Who are the individuals?

What are the variables of interest?

What is the explanatory variable?

What is the response variable?

Is this an experiment or an observational study?

7.   A researcher wants to see the relationship between sleep and performance among VCU students. An email was sent out to all current VCU students, asking them if they'd like to participate in the study. For those who replied, a survey was sent out, asking them to record the number of hours they slept over the course of two weeks. In addition, the students were to report their GPA.

What is the population of interest?

VCU students

What is the sampling frame?

All current VCU students

Who comprises the sample?

The students who agreed to participate

Who are the individuals?

Each student

What are the variables of interest?

Amount of sleep and GPA

What is the explanatory variable?

Amount of sleep

What is the response variable?

GPA

Is this an experiment or an observational study?

Observational study

8. A new pill has been created to ease the symptoms of migraine headaches. To test the effectiveness of this pill, 600 patients were randomly selected from the patient lists of doctors who consider themselves to be 'migraine specialists.' Two hundred of those patients were randomly selected to take the experimental drug, two hundred were randomly selected to get a placebo, and the rest stuck with their current plan. At the end of the month, the number of migraine headaches was compared for each group.

What is the population of interest?

What is the sampling frame?

Who comprises the sample?

Who are the individuals?

What are the variables of interest?

What is the explanatory variable?

What is the response variable?

Is this an experiment or an observational study?

8.   A new pill has been created to ease the symptoms of migraine headaches. To test the effectiveness of this pill, 600 patients were randomly selected from the patient lists of doctors who consider themselves to be 'migraine specialists.' Two hundred of those patients were randomly selected to take the experimental drug, two hundred were randomly selected to get a placebo, and the rest stuck with their current plan. At the end of the month, the number of migraine headaches was compared for each group.

What is the population of interest?
Migraine sufferers

What is the sampling frame?
People who see migraine specialists

Who comprises the sample?
The 600 selected people

Who are the individuals?
Each person with migraines

What are the variables of interest?
The pill choice and migraines

What is the explanatory variable?
Pill choice

What is the response variable?
Migraines

Is this an experiment or an observational study?
Experiment

Sample Collection Methods

Identify the sample collection method described in each situation.

A CEO of a department store wants to determine the level of satisfaction of his employees. He has fifteen stores throughout the central Virginia area.

1. He sends out an email, inviting employees to respond.

2. He randomly selects three stores, and uses every employee from those three stores.

3. He obtains an alphabetical list of all employees, using every tenth name on the list.

4. He randomly selects seven stores. From within each of those seven stores, he randomly selects ten employees each.

5. He uses every employee in his sample.

6. He lists all of the employees by code number. He uses software to generate code numbers.

7. He randomly selects six employees from every store.

8. He uses every employee from the three stores closest to his house.

A CEO of a department store wants to determine the level of satisfaction of his employees. He has fifteen stores throughout the central Virginia area.

1.   He sends out an email, inviting employees to respond.

Voluntary response

2.   He randomly selects three stores, and uses every employee from those three stores.

Cluster

3.   He obtains an alphabetical list of all employees, using every tenth name on the list.

Systematic

4.   He randomly selects seven stores. From within each of those seven stores, he randomly selects ten employees each.

Multistage

5.   He uses every employee in his sample.

Census

6.   He lists all of the employees by code number. He uses software to generate code numbers.

Simple random

7.   He randomly selects six employees from every store.

Stratified

8.   He uses every employee from the three stores closest to his house.

Convenience

9. To test the quality of spark plugs that come off of an assembly line, every twentieth spark plug was selected to be tested.

10. To survey VCU faculty on their opinion on a particular matter, six departments were chosen, and every faculty member from those six departments were used.

11. To test the quality of cucumbers in a grocery store chain, three cucumbers were randomly selected from every store in the district.

12. To survey employees at a large company, ten out of the thirty departments were randomly selected. From within each of those ten departments, five employees were randomly selected to participate.

13. To see how many beans are being produced by bean plants at farms in the county, twelve farms were randomly selected, and twenty plants were randomly selected from each of those twelve farms.

14. To determine if a professor was worthy of promotion, ten of his classes were randomly selected, and every student evaluation from those ten classes was used.

9. To test the quality of spark plugs that come off of an assembly line, every twentieth spark plug was selected to be tested.

Systematic

10. To survey VCU faculty on their opinion on a particular matter, six departments were chosen, and every faculty member from those six departments were used.

Cluster

11. To test the quality of cucumbers in a grocery store chain, three cucumbers were randomly selected from every store in the district.

Stratified

12. To survey employees at a large company, ten out of the thirty departments were randomly selected. From within each of those ten departments, five employees were randomly selected to participate.

Multistage

13. To see how many beans are being produced by bean plants at farms in the county, twelve farms were randomly selected, and twenty plants were randomly selected from each of those twelve farms.

Multistage

14. To determine if a professor was worthy of promotion, ten of his classes were randomly selected, and every student evaluation from those ten classes was used.

Cluster

Vocabulary (2)

1. It has been estimated that sixty million Americans suffer from allergies. To determine the effectiveness of an allergy pill, a randomly selected group of 1000 people took either the pill or the placebo (randomly selected.) The results were as follows: for indoor allergies, the pill was 18% effective; for outdoor allergies, the pill was 73% effective; for pet allergies, the pill was 51% effective.

Determine if the following are parameters or statistics:

1000

Sixty million

18%

Is bias present?

Is variability present?

2. A hotel chain had 65,672 guests last year. Surveys were mailed out to 1000 randomly selected guests; 50 people replied. The survey revealed that 78% of the guests were dissatisfied with the customer service.

Determine if the following are parameters or statistics:

65,672

1000

78%

Is bias present?

Is variability present?

1.   It has been estimated that sixty million Americans suffer from allergies. To determine the effectiveness of an allergy pill, a randomly selected group of 1000 people took either the pill or the placebo (randomly selected.) The results were as follows: for indoor allergies, the pill was 18% effective; for outdoor allergies, the pill was 73% effective; for pet allergies, the pill was 51% effective.

Determine if the following are parameters or statistics:

1000   Statistic

Sixty million   Parameter

18%   Statistic

Is bias present?   No

Is variability present? Yes, the answers are all over the place

2.   A hotel chain had 65,672 guests last year. Surveys were mailed out to 1000 randomly selected guests; 50 people replied. The survey revealed that 78% of the guests were dissatisfied with the customer service.

Determine if the following are parameters or statistics:

65,672   Parameter

1000   Statistic

78%   Statistic

Is bias present? Yes, voluntary response has negative bias

Is variability present? No

Margin of Error/Confidence Intervals

1. A random sample of 400 high school teachers revealed that 68% gave homework on Fridays. Calculate and interpret a 95% confidence interval for the proportion of high school teachers who give homework on Fridays.

What is p-hat?

What is the margin of error?

What is the interval?

Write the confidence statement.

2. A random sample of 300 phone customers showed that 145 had unlimited data plans. Calculate and interpret a 95% confidence interval for the proportion of all phone customers who have unlimited data.

What is p-hat?

What is the margin of error?

What is the interval?

Write the confidence statement.

1.   A random sample of 400 high school teachers revealed that 68% gave homework on Fridays. Calculate and interpret a 95% confidence interval for the proportion of high school teachers who give homework on Fridays.

What is $\hat{p}$?

$\hat{p}$= .68   (percent form)

What is the margin of error?

$$\frac{1}{\sqrt{n}} = \frac{1}{\sqrt{400}} = \frac{1}{20} = .05$$

What is the interval?

.68 - .05 = .63   .68 + .05 = .73   (.63,.73)

Write the confidence statement.

I am 95% confident that the proportion of all high school teachers who give homework on Fridays is between .63 and .73.

2.   A random sample of 300 phone customers showed that 145 had unlimited data plans. Calculate and interpret a 95% confidence interval for the proportion of all phone customers who have unlimited data.

What is p-hat?

145/300 = .4833

What is the margin of error?

$1/\sqrt{n} = 1/\sqrt{300}$ = 1/17.3205 = .0577

What is the interval?

.4833 - .0577 = .4256   .4833 + .0577 = .5410   (.4256, .5410)

Write the confidence statement.

I am 95% confident that the proportion of all phone customers who have unlimited data plans is between .4256 and .5410

3. A random sample of 256 VCU students showed that 84 had taken statistics. Calculate and interpret a 95% confidence interval for the proportion of all VCU students who have taken statistics.

What is p-hat?

What is the margin of error?

What is the interval?

Write the confidence statement.

4. A random sample of 600 VCU students revealed that 48% have taken the inter-campus connector. Calculate and interpret a 95% confidence interval for the proportion of all VCU students who have taken the inter-campus connector.

What is p-hat?

What is the margin of error?

What is the interval?

Write the confidence statement.

A random sample of 256 VCU students showed that 84 had taken statistics. Calculate and interpret a 95% confidence interval for the proportion of all VCU students who have taken statistics.

What is p-hat?

p = 84/256 = .3281

What is the margin of error?

$1/\sqrt{256}$ = $1/16$ = .0625

What is the interval?

.3281 - .0625 = .2656   .3281 + .0625 = .3906

Write the confidence statement.

I am 95% confident that the proportion of all VCU students who have taken statistics is between .2656 and .3906

A random sample of 600 VCU students revealed that 48% have taken the inter-campus connector. Calculate and interpret a 95% confidence interval for the proportion of all VCU students who have taken the inter-campus connector.

What is p-hat?

p = 0.48 (just move the decimal point since it was given as a percent)

What is the margin of error?

$1/\sqrt{600}$ = $1/(24.4949)$ = .0408

What is the interval?

.48 - .0408 = .4392   .48 + .0408 = .5208

Write the confidence statement.

I am 95% confident that the proportion of all VCU students who have taken the inter-campus connector is between .4392 and .5208

Questions to ask when reading about a study

Determine if the following are trustworthy sources:

1. A study conducted by a major toy maker has determined that children do better in school when they play games with their parents.

2. A major university discovered a relationship between Vitamin B6 and a lower incidence of a particular disease.

3. Doctors at a nationally known children's hospital claim that a treatment formerly reserved for cancer is also effective in certain auto-immune disorders.

4. A study conducted by Democrats revealed that 93% of people polled oppose the Republicans' new tax proposal.

1.   **A study conducted by a major toy maker has determined that children do better in school when they play games with their parents.**

No, the toy maker is trying to sell games/toys

2.   **A major university discovered a relationship between Vitamin B6 and a lower incidence of a particular disease.**

Universities are trustworthy sources

3.   **Doctors at a nationally known children's hospital claim that a treatment formerly reserved for cancer is also effective in certain auto-immune disorders.**

Doctors/hospitals are trustworthy sources

4.   **A study conducted by Democrats revealed that 93% of people polled oppose the Republicans' new tax proposal.**

No, Democrats want to sway people to their opinion

Questions to ask when reading about studies

Determine the problem with each of the following scenarios

1. A national maker of oral hygiene products conducted a study that showed flossing causes people to live longer.

2. A pill commercial advertises that the pill is twice as effective as the placebo.

3. A study done on college students revealed that people's reflexes start to slow after being awake for nineteen hours.

4. 80% of people recommend our product, which means only 20% recommend competitors' products.

5. People who switched to our insurance company saved an average of $300.

Questions to ask when reading about studies

Determine the problem with each of the following scenarios

1.   A national maker of oral hygiene products conducted a study that showed flossing causes people to live longer.

Not a trustworthy source; they are trying to sell products.

2.   A pill commercial advertises that the pill is twice as effective as the placebo.

"Twice as effective" as the placebo means it is probably not very effective.

3.   A study done on college students revealed that people's reflexes start to slow after being awake for nineteen hours.

College students don't represent the population

4.   80% of people recommend our product, which means only 20% recommend competitors' products.

These percentages don't need to add to make 100. It is possible that a person can recommend more than one product.

5.   People who switched to our insurance company saved an average of $300.

The sample is "people who switched." People would only switch if they were going to save money. This does not mean everybody would save an average of $300.

Determine which would be a better survey question:

1. Are you a charitable person?

Or

1. How much money have you given to charity in the past year?

2. What is your highest level of education?

Or

2. Are you well educated?

3. Are you a law abiding citizen?

Or

3. How many times have you been convicted of a crime?

4. Are you healthy?

Or

4. How many times did you go to the doctor for health-related issues last year?

Determine which would be a better survey question:

1.   Are you a charitable person?

Or

(1.)   How much money have you given to charity in the past year?

(2.)   What is your highest level of education?

Or

2. Are you well educated?

3. Are you a law abiding citizen?

Or

(3. How many times have you been convicted of a crime?

4. Are you healthy?

Or

(4. How many times did you go to the doctor for health-related issues last year?

Design of Experiments

A study was done to see the effectiveness of a new allergy pill. A random sample of 500 people with cat allergies was selected. Half were randomly selected to receive the pill, and the other half received a placebo. After three months, subjects were brought into a room with cats, and their symptoms were measured.

What is the treatment?

Who comprises the control group?

Is this single or double blind?

Can these results be extended to seasonal allergy sufferers?

Would confounding be a problem here?

Is this an experiment or an observational study?

Can cause and effect be declared?

A study was done to see the effectiveness of a new allergy pill. A random sample of 500 people with cat allergies was selected. Half were randomly selected to receive the pill, and the other half received a placebo. After three months, subjects were brought into a room with cats, and their symptoms were measured.

What is the treatment?

The allergy pill

Who comprises the control group?

People taking the placebo

Is this single or double blind?

Double blind: you can always assume the people measuring don't know group assignment. In this case, the subjects would also not know which pill they took.

Can these results be extended to seasonal allergy sufferers?

No, a separate study would need to be done for seasonal allergy sufferers

Would confounding be a problem here?

No, the groups were assigned randomly

Is this an experiment or an observational study?

Experiment

Can cause and effect be declared?

Yes, you may declare cause and effect at the end of an experiment

A study was done to see the effectiveness of store-brand sunblock compared to name-brand sunblock. 100 randomly selected beachgoers put the store brand on the one side of their bodies and put the name-brand on the other side of their bodies. Subjects were asked to reapply equally every two hours. Subjects were also asked to keep track of how much time they spent in the water.

Is this an experiment or an observational study?

What is the treatment?

What interacting variable has been taken into account?

Can the results of this study be extended?

If subjects were divided into three categories depending on skin tone (light, medium, dark) and the results were investigated separately, what phenomenon would that be?

A study was done to see the effectiveness of store-brand sunblock compared to name-brand sunblock. 100 randomly selected beachgoers put the store brand on the one side of their bodies and put the name-brand on the other side of their bodies. Subjects were asked to reapply equally every two hours. Subjects were also asked to keep track of how much time they spent in the water.

Is this an experiment or an observational study?

Experiment; treatments were deliberately imposed

What is the treatment?

Sunblock

What interacting variable has been taken into account?

Time in the water

Can the results of this study be extended?

Yes, the subjects were randomly selected, sunblock is realistic and it was done in a natural setting

If subjects were divided into three categories depending on skin tone (light, medium, dark) and the results were investigated separately, what phenomenon would that be?

Blocking

A researcher wanted to do a study to determine the opinions of young adults regarding global warming. He went to a local university to obtain a randomly selected sample of 200 students, asking about their feelings toward global warming.

Is this an experiment or an observational study?

Can the results of this study be extended to the target population?

Explain how the Hawthorne Effect can be a problem in this situation: your boss wants to see how well you are adhering to the guidelines of the company, so he tells you he will be observing you next Wednesday from 11 to 12.

Explain how confounding can be a problem here: A teacher has posted a practice test online, and students can choose whether or not to take it. The teacher has the ability to see which students have completed the practice test. After the exam, the teacher compares the averages of students who took the pretest to students who didn't.

A researcher wanted to do a study to determine the opinions of young adults regarding global warming. He went to a local university to obtain a randomly selected sample of 200 students, asking about their feelings toward global warming.

Is this an experiment or an observational study?

Observational study

Can the results of this study be extended to the target population?

No, only young people who go to college are in the study. Young people who do not go to college aren't considered. (The sample doesn't represent the whole population)

Explain how the Hawthorne Effect can be a problem in this situation: your boss wants to see how well you are adhering to the guidelines of the company, so he tells you he will be observing you next Wednesday from 11 to 12.

When employees know they are being observed, they are more likely to act in accordance with the company guidelines. Knowing in advance that the observation is coming can make the employee even more compliant.

Explain how confounding can be a problem here: A teacher has posted a practice test online, and students can choose whether or not to take it. The teacher has the ability to see which students have completed the practice test. After the exam, the teacher compares the averages of students who took the pretest to students who didn't.

Practice Test	No Practice Test
Studious kids take the practice test	Students who care less about their grades don't take the practice test
OR students who struggle take the practice test	Students who understand the material don't take the practice test

*Students should have been randomly selected to take the practice test

Determining the soundness of arguments:

If there are flaws in the following arguments, identify them.

1. The number of suicides has increased dramatically over the past 50 years; therefore, suicide is more of a problem today.

2. I measured the temperatures of three cooked turkeys, and none of them had reached the desired temperature. I used the same meat thermometer on all three, so I know it's correct and the ovens must not be working properly.

3. The test used to determine kindergarten readiness asks children to point to such body parts as their hip and waist, as well as recite their zip code. This test adequately measures how intelligent the child is.

4. 60% of customers from company A would recommend A to a friend. 92% of customers from company B would recommend B to a friend. Therefore, I should go with company B.

5. More people have cell phones than ever before. More people have brain cancer than ever before. Cell phones must be causing brain cancer.

6. The company's website offers customer testimonials, which are all positive. Therefore, this company must be good.

Determining the soundness of arguments:

If there are flaws in the following arguments, identify them.

1. The number of suicides has increased dramatically over the past 50 years; therefore, suicide is more of a problem today.

The number of suicides has increased, but so has the population. One must look at the percent of people committing suicide. (Rate vs. count)

2. I measured the temperatures of three cooked turkeys, and none of them had reached the desired temperature. I used the same meat thermometer on all three, so I know it's correct and the ovens must not be working properly.

The thermometer might be running low (bias)

3. The test used to determine kindergarten readiness asks children to point to such body parts as their hip and waist, as well as recite their zip code. This test adequately measures how intelligent the child is.

The test only identifies whether an adult has taught the child those particular terms. It does not measure intelligence. (validity issue)

4. 60% of customers from company A would recommend A to a friend. 92% of customers from company B would recommend B to a friend. Therefore, I should go with company B.

Since rates are used and the numbers are substantially different, this argument is sound.

5. More people have cell phones than ever before. More people have brain cancer than ever before. Cell phones must be causing brain cancer.

An increase in the population would explain both.

6. The company's website offers customer testimonials, which are all positive. Therefore, this company must be good.

The company is in control of what it puts on their website; they would not put negative testimonials. An objective, third-party review would be better (bias.)

Correlation/Regression examples

The following table relates the square footage of local apartments and the monthly rent.

Square Feet	Monthly Rent
724	870
803	950
815	980
847	1000
876	995
906	1100
925	1125
955	1120
979	1200
980	1150
1023	1225
1056	1250
1088	1300
1122	1325
1169	1375

What is the value of the correlation coefficient r? Why is it positive?

Would the regression equation be a good tool for predicting rent based on square footage?

What is the regression equation?

What would be the predicted rent for an apartment that is 1000 square feet?

Correlation/Regression examples

The following table relates the square footage of local apartments and the monthly rent.

Square Feet	Monthly Rent
724	870
803	950
815	980
847	1000
876	995
906	1100
925	1125
955	1120
979	1200
980	1150
1023	1225
1056	1250
1088	1300
1122	1325
1169	1375

What is the value of the correlation coefficient r? Why is it positive?

.9908   As square footage goes up, rent goes up

Would the regression equation be a good tool for predicting rent based on square footage?

Yes, the correlation coefficient is close to one

What is the regression equation?

Y = 15.7425 + 1.1725 X   Y = rent,   X = square footage

What would be the predicted rent for an apartment that is 1000 square feet?

$1188.2425

The following table relates the number credits a student is taking compared to the number of hours worked per week (among working students).

Number of credits	Number of hours worked
3	40
3	36
3	40
4	40
6	35
6	32
7	35
9	20
10	22
12	30
12	10
13	15
14	20
14	25
15	22
15	12
17	8
18	10

What is the correlation coefficient? Why is this value negative?

Is there a strong relationship between credits and hours worked?

What is the regression equation?

What would be the predicted number of hours worked for a student taking 9 credits? (PREDICTED, not observed)

The following table relates the number credits a student is taking compared to the number of hours worked per week (among working students).

Number of credits	Number of hours worked
3	40
3	36
3	40
4	40
6	35
6	32
7	35
9	20
10	22
12	30
12	10
13	15
14	20
14	25
15	22
15	12
17	8
18	10

What is the correlation coefficient? Why is this value negative?

-.8872   As the number of credits a student is taking goes up, the number of hours worked goes down

Is there a strong relationship between credits and hours worked?

Relatively strong...the correlation coefficient value is close to one

What is the regression equation?

$Y = 44.8821 - 1.9662 X$ Y = hours worked per week, X = credit hours

What would be the predicted number of hours worked for a student taking 9 credits? (PREDICTED, not observed)

27.1863 hours

The following table relates the heights and weights of a random sample of students

Height in inches	Weight in pounds
60	150
70	185
65	172
68	219
72	199
74	225
62	225
67	155
65	183
70	160
71	187
66	207

What is the value of the correlation coefficient? Why is this value positive?

What is the regression equation?

What would be the predicted weight of someone who is 63 inches tall?

Is this a reliable prediction?

Using this information, predict the weight of a newborn who is 21 inches long (tall). Why is this number unreasonable?

The following table relates the heights and weights of a random sample of students

Height in inches	Weight in pounds
60	150
70	185
65	172
68	219
72	199
74	225
62	225
67	155
65	183
70	160
71	187
66	207

What is the value of the correlation coefficient? Why is this value positive?

.2858   As height goes up, weight goes up.

What is the regression equation?

Y = 65.1667 + 1.8333 X   Y = weight in pounds, X = height in inches

What would be the predicted weight of someone who is 63 inches tall?

180.6646

Is this a reliable prediction?

No, the correlation coefficient is closer to zero than one.

Using this information, predict the weight of a newborn who is 21 inches long (tall). Why is this number unreasonable?

103.666 pounds. This value is an extrapolation since it is outside the given range of X's.

Which type of graph is appropriate?

Which type(s) of graphs would be appropriate? The choices are line graph, bar graph, histogram and pie chart.

1. You wish to show how many students have each type of pet.

2. You wish to track your weight loss over the course of the year.

3. You wish to display the income brackets of people at a company.

4. You wish to show the class status (freshman, sophomore...) of your classmates.

5. You wish to display the GPAs of your classmates.

6. You wish to track your number of steps taken each day for a week.

7. You wish to display how many students belong to each club at school.

Which type of graph is appropriate?

Which type(s) of graphs would be appropriate? The choices are line graph, bar graph, histogram and pie chart.

1. You wish to show how many students have each type of pet.

Bar graph; categorical but people may fall into more than one category.

2. You wish to track your weight loss over the course of the year.

Line graph; change over time

3. You wish to display the income brackets of people at a company.

Histogram; numerical data

4. You wish to show the class status (freshman, sophomore...) of your classmates.

Bar graph or pie chart; categorical, and you asked one group of people one question with all options listed. Nobody can be in more than one group.

5. You wish to display the GPAs of your classmates.

Histogram; numerical data

6. You wish to track your number of steps taken each day for a week.

Line graph; change over time

7. You wish to display how many students belong to each club at school.

Bar graph; categorical, but people may belong in more than one group.

Identify the following statements as true or false:

Line graphs should be used to display categorical data.

Bar graphs have spaces between the bars.

Histograms display quantitative (numerical) data.

Pie charts either need a key or should have the sections labeled.

Histograms and bar graphs should have axis labels.

Histograms have spaces between the bars.

All graphs should have a title.

The Y-axis should always begin at zero.

The X-axis indicates whether the information is categorical or numerical.

Bar graphs can be used whenever the data is categorical, even if someone can fall into more than one category.

Pie charts can be used when you ask one group of people one question and all options are listed—and no one can be in more than one group.

Line graphs should be used to display categorical data.

False, line graphs show change over time

Bar graphs have spaces between the bars.

True

Histograms display quantitative (numerical) data.

True

Pie charts either need a key or should have the sections labeled.

True

Histograms and bar graphs should have axis labels.

True

Histograms have spaces between the bars.

False, histograms have bars that connect

All graphs should have a title.

True

The Y-axis should always begin at zero.

True

The X-axis indicates whether the information is categorical or numerical.

True

Bar graphs can be used whenever the data is categorical, even if someone can fall into more than one category.

True

Pie charts can be used when you ask one group of people one question and all options are listed—and no one can be in more than one group.

True

Summarizing Data

Determine the shape of the following data set, and then determine the appropriate measure of center and spread. (Calculate those values.)

20	21	21	22	22	24	25	27	27	29	29	30	33
35	56											

Shape: **Center:**   **Spread:**

64	67	69	69	70	71	72	73	73	74	74	74	75
75	76	77	78	79	79	81	81	84				

Shape: **Center:**   **Spread:**

Summarizing Data

Determine the shape of the following data set, and then determine the appropriate measure of center and spread. (Calculate those values.)

20	21	21	22	22	24	25	27	27	29	29	30	33
35	56											

Shape:

Skewed to the right

Center:

Median = 27

Spread:

5 number summary = Min = 20

Q1 = 22, Med = 27, Q3 = 30, Max = 56

64	67	69	69	70	71	72	73	73	74	74	74	75
75	76	77	78	79	79	81	81	84				

Shape:

Bell

Center:

Mean = 74.3182

Spread:

Standard deviation = 4.9124

The amount of wait time, in minutes, at the Department of Motor Vehicles was recorded for a random sample of 15 people.

20   24   25   29   33   35   36   40   42   44   46   48   55

56   78

Summarize the data using the best approach for the shape.

Shape: Center: Spread:

Grades from a particular exam are as follows:

59   63   65   68   70   73   75   76   77   77   79   81   85

86   88   90   93   96

Summarize the data using the best approach for the shape.

Shape: Center: Spread:

The amount of wait time, in minutes, at the Department of Motor Vehicles was recorded for a random sample of 15 people.

20   24   25   29   33   35   36   40   42   44   46   48   55

56   78

Summarize the data using the best approach for the shape.

Shape: Center: Spread:

Skewed to the right   **Median = 40**   **Min = 20, Q1 = 29, Med = 40,**

Q3 = 48, Max = 78

Grades from a particular exam are as follows:

59   63   65   68   70   73   75   76   77   77   79   81   85

86   88   90   93   96

Summarize the data using the best approach for the shape.

Shape: Center: Spread:

Bell   **Mean = 77.8333**   **Standard Deviation = 10.5007**

The heights of 17 randomly selected students were recorded and are as follows:

| 64 | 69 | 59 | 63 | 68 | 74 | 64 | 62 | 67 | 67 | 67 | 69 | 67 |
| 67 | 65 | 70 | 70 |

Shape: Center: Spread:

The number of books read for pleasure (last year) from a random sample of 30 students is as follows:

5	6	50	0	2	39	1	3	9	10	4	9	5
9	10	11	40	5	4	5	0	1	1	2	3	8
11	12	55	10

Shape: Center: Spread:

The heights of 17 randomly selected students were recorded and are as follows:

64 69 59 63 68 74 64 62 67 67 67 69 67

67 65 70 70

Shape: **Center:**   **Spread:**

Bell   Mean = 66.5882   Standard deviation = 3.5366

The number of books read for pleasure (last year) from a random sample of 30 students is as follows:

5   6   50 0   2   39 1   3   9   10 4   9   5

9   10 11 40 5   4   5   0   1   1   2   3   8

11 12 55 10

Shape: **Center:**   **Spread:**

Skewed to the right   Median = 5.5   Min = 0, Q1 = 3, Med = 5.5

                                                            Q3 = 10, Max = 55

Grades from a recent exam are as follows:

0   68   70   72   72   74   76   77   77   78   80   82   84

85   88   89   90   92   97

Shape: Center: Spread:

When do you use mean?

When do you use median?

Which value from the calculator do you use for standard deviation?

Grades from a recent exam are as follows:

0   68   70   72   72   74   76   77   77   78   80   82   84

85   88   89   90   92   97

Shape: Center: Spread:

Skewed to the left   **Med = 78**   **Min = 0, Q1 = 72, Med = 78, Q3 = 88,**

Max = 97

When do you use mean?

When the curve is bell shaped/free of outliers

When do you use median?

When the data has outliers

Which value from the calculator do you use for standard deviation?

Always S_x

Draw and label normal curves

68-95-99.7 rule

Draw and label a normal curve with a mean of 22 and standard deviation of 5.

68% of the values fall between...

99.7% of the values fall between...

IQs are normally distributed with a mean of 100 and a standard deviation of 15. Draw the curve.

What percent of people have an IQ between 70 and 130?

What percent of people have an IQ between 85 and 115?

Draw and label normal curves

68-95-99.7 rule

Draw and label a normal curve with a mean of 22 and standard deviation of 5.

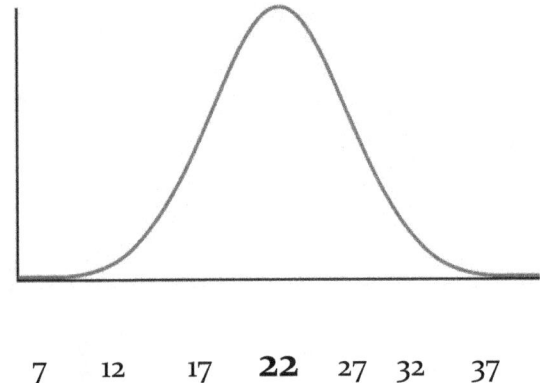

7   12   17   **22**   27   32   37

68% of the values fall between... **17 and 27**

99.7% of the values fall between...   **7 and 37**

IQs are normally distributed with a mean of 100 and a standard deviation of 15. Draw the curve.

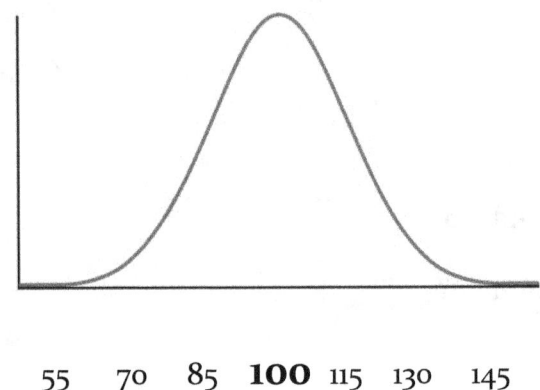

55   70   85   **100**   115   130   145

What percent of people have an IQ between 70 and 130? **95%**

What percent of people have an IQ between 85 and 115? **68%**

Z-score Transformations

IQs are normally distributed with a mean of 100 and a standard deviation of 15. Use that information to complete the following:

Adrienne's IQ is 113. What is her percentile?

What IQ is at the 43rd percentile?

Dave's IQ is 84. What percent of people have an IQ higher than him?

Sue's IQ is 122. What percent of people have an IQ higher than her?

What IQ is at the 25th percentile?

What IQ is at the 80th percentile?

Z-score Transformations

IQs are normally distributed with a mean of 100 and a standard deviation of 15. Use that information to complete the following:

Adrienne's IQ is 113. What is her percentile?

$\mu = 100$ $\sigma = 15$ $\bar{x} = 113$ $n = 1$ *direction:* $<$ .8069 = 80.69%

What IQ is at the 43rd percentile?

area $= .43$ $\mu = 100$ $\sigma = 15$ 97.3544

Dave's IQ is 84. What percent of people have an IQ higher than him?

$\mu = 100$ $\sigma = 15$ $\bar{x} = 84$ $n = 1$ *direction:* $>$ .8569 = 85.69%

Sue's IQ is 122. What percent of people have an IQ higher than her?

$\mu = 100$ $\sigma = 15$ $\bar{x} = 122$ $n = 1$ *direction:* $>$ .0712 = 7.12%

What IQ is at the 25th percentile?

area $= .25$ $\mu = 100$ $\sigma = 15$ 89.8827

What IQ is at the 80th percentile?

area $= .80$ $\mu = 100$ $\sigma = 15$ 112.624

Z-score Transformations

1. Adult cat lengths are normally distributed with a mean of 18 inches (without tail) and a standard deviation of 1.9 inches. What percent of cats are more than 19 inches long?

2. Heights of men are normally distributed with a mean of 69.25 inches and a standard deviation of 3.1 inches. What percent of men are less than 68 inches tall?

3. Once diagnosed with Cushing's Disease, the amount of time that dogs live is normally distributed with a mean of 3.1 years with a standard deviation of 1.2 years. If a dog is at the 75th percentile, how long did that dog live after diagnosis?

4. Times to complete a race are normally distributed with a mean of 32.6 minutes and a standard deviation of 5.4 minutes. If you finished in 41.3 minutes, what percent of people had a time lower than you?

5. SAT scores are normally distributed with a mean of 1600 and a standard deviation of 300. If you score at the 70th percentile, what was your score?

Z-score Transformations

1.   Adult cat lengths are normally distributed with a mean of 18 inches (without tail) and a standard deviation of 1.9 inches. What percent of cats are more than 19 inches long?

$\mu = 18$, $\sigma = 1.9$ $\bar{x} = 19$ $n = 1$ $direction:$ >   **.2993 = 29.93%**

2.   Heights of men are normally distributed with a mean of 69.25 inches and a standard deviation of 3.1 inches. What percent of men are less than 68 inches tall?

$\mu = 69.25$, $\sigma = 3.1$ $\bar{x} = 68$ $n = 1$ $direction:$ <   **.3434 = 34.34%**

3.   Once diagnosed with Cushing's Disease, the amount of time that dogs live is normally distributed with a mean of 3.1 years with a standard deviation of 1.2 years. If a dog is at the 75th percentile, how long did that dog live after diagnosis?

$Area = .75$ $\mu = 3.1$ $\sigma = 1.2$ **3.9094 years**

4.   Times to complete a race are normally distributed with a mean of 32.6 minutes and a standard deviation of 5.4 minutes. If you finished in 41.3 minutes, what percent of people had a time lower than you?

$\mu = 32.6$, $\sigma = 5.4$ $\bar{x} = 41.3$ $n = 1$ $direction:$ <   **.9464 = 94.64%**

5.   SAT scores are normally distributed with a mean of 1600 and a standard deviation of 300. If you score at the 70th percentile, what was your score?

$Area = .70$ $\mu = 1600$ $\sigma = 300$ **1757.32**

Z-score Transformations

1. Wait times at an emergency room are normally distributed with a mean of 17 minutes and a standard deviation of 6 minutes. If you wait time fell at the 60th percentile, how long did you wait?

2. Scores on a college entrance exam are normally distributed with a mean of 300 and a standard deviation of 15. If you scored a 306, what was your percentile?

3. The number of hits a website receives each day is normally distributed with a mean of 160 and a standard deviation of 18. Tuesday, they were at the 30th percentile. How many hits did they get?

4. The ages when women have their first child are normally distributed with a mean of 28.6 years and a standard deviation of 5.1 years. If Susan was 31 when she had her first child, what percent of women were older than her when they had their first child? (What percent of women had a higher age than Susan?)

5. The number of people who sleep at a homeless shelter is normally distributed with a mean of 122.2 and a standard deviation of 15.4. On a particularly cold night, 151 people slept at the shelter. What percent of the time do fewer people sleep there than that?

Z-score Transformations

1. Wait times at an emergency room are normally distributed with a mean of 17 minutes and a standard deviation of 6 minutes. If you wait time fell at the 60th percentile, how long did you wait?

   $Area = .60$ $\mu = 17$ $\sigma = 6$ **18.5201 minutes**

2. Scores on a college entrance exam are normally distributed with a mean of 300 and a standard deviation of 15. If you scored a 306, what was your percentile?

   $\mu = 300, \sigma = 15$ $\bar{x} = 306$ $n = 1$ $direction:$ $<$ **.6554 = 65.54%**

3. The number of hits a website receives each day is normally distributed with a mean of 160 and a standard deviation of 18. Tuesday, they were at the 30th percentile. How many hits did they get?

   $Area = .30$ $\mu = 160$ $\sigma = 18$ **150.5608 hits**

4. The ages when women have their first child are normally distributed with a mean of 28.6 years and a standard deviation of 5.1 years. If Susan was 31 when she had her first child, what percent of women were older than her when they had their first child? (What percent of women had a higher age than Susan?)

   $\mu = 28.6, \sigma = 5.1$ $\bar{x} = 31$ $n = 1$ $direction:$ $>$ **.3189 = 31.89%**

5. The number of people who sleep at a homeless shelter is normally distributed with a mean of 122.2 and a standard deviation of 15.4. On a particularly cold night, 151 people slept at the shelter. What percent of the time do fewer people sleep there than that?

   $\mu = 122.2, \sigma = 15.4$ $\bar{x} = 151$ $n = 1$ $direction:$ $<$ **.9693 = 96.93%**

Averaging groups together

1. Weight losses from a particular diet plan are normally distributed with a mean of 45.6 pounds and a standard deviation of 11.8 pound. What is the probability that a randomly selected group of 37 people would have an average weight loss less than 44 pounds?

2. Grades from a recent exam are normally distributed with a mean of 78.8 and a standard deviation of 4.7. What is the probability that a randomly selected group of 30 people have an average grade higher than 80?

3. Summer high temperatures for a particular city are normally distributed with a mean of 91.8 degrees and a standard deviation of 3.8 degrees. What is the probably that a randomly selected group of 10 days will have an average temperature above 90 degrees?

4. The lengths of naps among ten-month-olds is normally distributed with a mean of 106.24 minutes and a standard deviation of 25.86 minutes. If 17 ten-month-olds were randomly selected, what is the probability their average nap time will be longer than 100 minutes?

Averaging groups together

1. Weight losses from a particular diet plan are normally distributed with a mean of 45.6 pounds and a standard deviation of 11.8 pound. What is the probability that a randomly selected group of 37 people would have an average weight loss less than 44 pounds?

$\mu = 45.6$, $\sigma = 11.8$ $\bar{x} = 44$ $n = 37$ *direction:* <   .2047 = 20.47%

2. Grades from a recent exam are normally distributed with a mean of 78.8 and a standard deviation of 4.7. What is the probability that a randomly selected group of 30 people have an average grade higher than 80?

$\mu = 78.8$, $\sigma = 4.7$ $\bar{x} = 80$ $n = 30$ *direction:* >   .0810 = 8.10%

3. Summer high temperatures for a particular city are normally distributed with a mean of 91.8 degrees and a standard deviation of 3.8 degrees. What is the probably that a randomly selected group of 10 days will have an average temperature above 90 degrees?

$\mu = 91.8$, $\sigma = 3.8$ $\bar{x} = 90$ $n = 10$ *direction:* >   .9329 = 93.29%

4. The lengths of naps among ten-month-olds is normally distributed with a mean of 106.24 minutes and a standard deviation of 25.86 minutes. If 17 ten-month-olds were randomly selected, what is the probability their average nap time will be longer than 100 minutes?

$\mu = 106.24$, $\sigma = 25.86$ $\bar{x} = 100$ $n = 17$ *direction:* >   .8401 = 84.01%

Mixed Examples (Z-score transformations, individuals and groups)

1. Male heights are normally distributed with a mean of 69.25 inches and a standard deviation of 2.9 inches. What is the probability that a randomly selected male is taller than 72 inches?

2. Female heights are normally distributed with a mean of 54 inches and a standard deviation of 2.7 inches. If a woman is at the 75th percentile, how tall is she?

3. One-year-old baby weights are normally distributed with a mean of 20.3 pounds and a standard deviation of 1.6 pounds. If a group of 100 babies is randomly selected, what is the probability that their average weight is less than 20 pounds?

4. Newborn baby lengths are normally distributed with a mean of 20.6 inches and a standard deviation of 1.8 inches. What baby length is at the 20th percentile?

5. Newborn baby weights are normally distributed with a mean of 7.2 pounds and a standard deviation of 2.1 pounds. What is the probability that a randomly selected group of 30 babies will have an average weight more than 7 pounds?

6. Full-grown Newfoundland dog weights are normally distributed with a mean of 178.4 pounds and a standard deviation of 12.8 pounds. What is the probability that a randomly selected Newfoundland weighs less than 150 pounds?

Mixed Examples (Z-score transformations, individuals and groups)

1. Male heights are normally distributed with a mean of 69.25 inches and a standard deviation of 2.9 inches. What is the probability that a randomly selected male is taller than 72 inches?

$\mu = 69.25$, $\sigma = 2.9$ $\bar{x} = 72$ $n = 1$ direction: $>$ **.1715 = 17.15%**

2. Female heights are normally distributed with a mean of 54 inches and a standard deviation of 2.7 inches. If a woman is at the 75th percentile, how tall is she?

Area = .75 $\mu = 54$ $\sigma = 2.7$ **55.8211 inches**

3. One-year-old baby weights are normally distributed with a mean of 20.3 pounds and a standard deviation of 1.6 pounds. If a group of 100 babies is randomly selected, what is the probability that their average weight is less than 20 pounds?

$\mu = 20.3$, $\sigma = 1.6$ $\bar{x} = 20$ $n = 100$ direction: $<$ **.0304 = 3.04%**

4. Newborn baby lengths are normally distributed with a mean of 20.6 inches and a standard deviation of 1.8 inches. What baby length is at the 20th percentile?

Area = .20 $\mu = 20.6$ $\sigma = 1.8$ **19.0851 inches**

5. Newborn baby weights are normally distributed with a mean of 7.2 pounds and a standard deviation of 2.1 pounds. What is the probability that a randomly selected group of 30 babies will have an average weight more than 7 pounds?

$\mu = 7.2$, $\sigma = 2.1$ $\bar{x} = 7$ $n = 30$ direction: $>$ **.6990 = 69.90 %**

6. Full-grown Newfoundland dog weights are normally distributed with a mean of 178.4 pounds and a standard deviation of 12.8 pounds. What is the probability that a randomly selected Newfoundland weighs less than 150 pounds?

$\mu = 178.4$, $\sigma = 12.8$ $\bar{x} = 150$ $n = 1$ direction: $<$ **.0133 = 1.33 %**

Mutually Exclusive and Independent

A certain population was polled, and here were partial results:

Belongs to a gym: 34%

Has children: 27%

Has no dogs: 56%

Has one dog: 37%

Has a car payment: 73%

1. If a person is selected at random, what is the probability that person belongs to a gym and has children?

2. If you select a person at random, what is the probability that person has no dogs or has one dog?

3. If you select a person at random, what is the probability that person has two or more dogs?

4. If you select two people at random, with replacement, what is the probability that both of them have car payments?

5. If you select one person at random, what is the probability that person has one dog and a car payment?

Mutually Exclusive and Independent

A certain population was polled, and here were partial results:

Belongs to a gym: 34%

Has children: 27%

Has no dogs: 56%

Has one dog: 37%

Has a car payment: 73%

1. If a person is selected at random, what is the probability that person belongs to a gym and has children?

$(.34)(.27) = .0918 = 9.18\%$

2. If you select a person at random, what is the probability that person has no dogs or has one dog?

$56\% + 37\% = 93\%$

3. If you select a person at random, what is the probability that person has two or more dogs?

$100\% - 93\% = 7\%$

4. If you select two people at random, with replacement, what is the probability that both of them have car payments?

$(.73)(.73) = .5329 = 53.29\%$

5. If you select one person at random, what is the probability that person has one dog and a car payment?

$(.37)(.73) = .2701 = 27.01\%$

Another population was polled, and here were the results:

Phone carrier "A": 26.3%

Phone carrier "B": 32.7%

Blue eyes: 28.9%

Brown eyes: 54.6%

1. If you pick a person at random, what is the probability that person uses a phone carrier other than A or B?

2. If you pick a person at random, what is the probability that person has blue eyes and uses phone carrier B?

3. If you pick two people at random, with replacement, what is the probability they both have brown eyes?

4. If you pick a person at random, what is the probability they have blue or brown eyes?

5. If you pick a person at random, what is the probability they use carrier A and have blue eyes?

Another population was polled, and here were the results:

Phone carrier "A": 26.3%

Phone carrier "B": 32.7%

Blue eyes: 28.9%

Brown eyes: 54.6%

1. If you pick a person at random, what is the probability that person uses a phone carrier other than A or B?

26.3% + 32.7% = 59%   100% - 59% = 41%

2. If you pick a person at random, what is the probability that person has blue eyes and uses phone carrier B?

(.289)(.327) = .0945 = 9.45%

3. If you pick two people at random, with replacement, what is the probability they both have brown eyes?

(.546)(.546) = .2981 = 29.81%

4. If you pick a person at random, what is the probability they have blue or brown eyes?

28.9% + 54.6% = 83.5%

5. If you pick a person at random, what is the probability they use carrier A and have blue eyes?

(.263)(.289) = .0760 = 7.60%

Relative Risk

1. The risk of catching the flu without a flu shot is 12.6%. The risk of getting the flu with the flu shot is 1.9%. What is the relative risk of getting the flu without the shot compared to with the shot?

2. The risk of getting pregnant on the birth control pill is 1%. The risk of getting pregnant with an IUD is 0.2%. What is the relative risk of getting pregnant on the birth control pill compared to an IUD?

3. The risk of having a baby with Down's syndrome is 11% when the mother is age 40. The risk of having a baby with Down's syndrome is 0.5% when the mother is age 35. What is the relative risk of having a baby with Down's syndrome at age 40 compared to age 35?

4. The risk of getting into an accident with an intoxicated driver is 7.4%. The risk of getting into an accident with a sober driver is 0.03%. What is the relative risk of getting into an accident with a drunk driver compared to a sober driver?

Relative Risk

1. The risk of catching the flu without a flu shot is 12.6%. The risk of getting the flu with the flu shot is 1.9%. What is the relative risk of getting the flu without the shot compared to with the shot?

12.6% / 1.9% = **6.6316 times as likely**

2. The risk of getting pregnant on the birth control pill is 1%. The risk of getting pregnant with an IUD is 0.2%. What is the relative risk of getting pregnant on the birth control pill compared to an IUD?

1% / 0.2% = **5 times as likely**

3. The risk of having a baby with Down's syndrome is 11% when the mother is age 40. The risk of having a baby with Down's syndrome is 0.5% when the mother is age 35. What is the relative risk of having a baby with Down's syndrome at age 40 compared to age 35?

11% / 0.5% = **22 times as likely**

4. The risk of getting into an accident with an intoxicated driver is 7.4%. The risk of getting into an accident with a sober driver is 0.03%. What is the relative risk of getting into an accident with a drunk driver compared to a sober driver?

7.4% / 0.03% = **246.667 times as likely**

Confidence Intervals on Proportions

1. A random sample of 56 VCU students revealed that 13 have vacationed outside of the United States. Calculate and interpret a 95% confidence interval for the proportion of all VCU students who have vacationed outside of the United States.

2. A random sample of 200 convicted criminals revealed a 64% recidivism rate (64% reoffended.) Calculate and interpret a 90% confidence interval for the percent of all convicted criminals who reoffend.

3. In a recent study, 72 out of 400 randomly selected student athletes had been to the doctor for a sports-related injury. Calculate and interpret a 99% confidence interval for the proportion of all student athletes who have been to the doctor for a sports-related injury.

4. A random sample of 300 VCU students revealed that 32% took an art class as an elective in high school. Calculate and interpret a 95% confidence interval for the proportion of all VCU students who took an art class in high school.

Confidence Intervals on Proportions

1. A random sample of 56 VCU students revealed that 13 have vacationed outside of the United States. Calculate and interpret a 95% confidence interval for the proportion of all VCU students who have vacationed outside of the United States.

X = 13   n = 56   c-level = .95   (.1216, .3427)

I am 95% confident that the proportion of all VCU students who have vacationed outside of the United States is between .1216 and .3417.

2. A random sample of 200 convicted criminals revealed a 64% recidivism rate (24% reoffended.) Calculate and interpret a 90% confidence interval for the proportion of all convicted criminals who reoffend.

64% of 200 criminals = (0.64)(200) = 128 criminals

x = 128   n = 200   c-level = .90   (.5842, .6958)

I am 90% confident that the proportion of all convicted criminals who reoffend is between .5842 and .6958.

3. In a recent study, 72 out of 400 randomly selected student athletes had been to the doctor for a sports-related injury. Calculate and interpret a 99% confidence interval for the proportion of all student athletes who have been to the doctor for a sports-related injury.

X = 72   n = 400   c-level = .99   (.1305, .2295)

I am 99% confident that the proportion of all student athletes who have been to the doctor for a sports-related injury is between .1305 and .2295.

4. A random sample of 300 VCU students revealed that 32% took an art class as an elective in high school. Calculate and interpret a 95% confidence interval for the proportion of all VCU students who took an art class in high school.

32% of 300 students = (.32)(300) = 96 students

X = 96   n = 300   c-level = .95   (.2672, .3728)

I am 95% confident that the proportion of VCU students who took an art class in high school is between .2672 and .3728.

Confidence Intervals on Means

1. A random sample of 200 Americans consumed an average of 42 grams of sugar per day in drinks, with a standard deviation of 14 grams. Calculate and interpret a 95% confidence interval for the average number of sugar grams consumed by all Americans in their drinks.

2. 500 randomly selected car owners drove an average of 17,384 miles per year, with a standard deviation of 4,688 miles. Calculate and interpret a 90% confidence interval for the average amount of miles driven each year.

3. According to a poll of 100 randomly selected Americans, the average number of days spent on vacation is 5.6 with a standard deviation of 1.2. Calculate and interpret a 99% confidence interval for the average number of days Americans spend on vacation.

4. A random sample of 400 Americans revealed the average amount of water consumed per day is 32.5 ounces, with a standard deviation of 10.6 ounces. Calculate and interpret a 90% confidence interval for the average number of ounces of water consumed per day by Americans.

Confidence Intervals on Means

1. A random sample of 200 Americans consumed an average of 42 grams of sugar per day in drinks, with a standard deviation of 14 grams. Calculate and interpret a 95% confidence interval for the average number of sugar grams consumed by all Americans in their drinks.

   $\sigma = 14$ $\bar{x} = 42$ $n = 200$ $c - level = .95$ $(40.0597, 43.9403)$

   I am 95% confident that the average number of grams of sugar consumed by Americans per day is between 40.0597 and 43.9403 grams.

2. 500 randomly selected car owners drove an average of 17,384 miles per year, with a standard deviation of 4,688 miles. Calculate and interpret a 90% confidence interval for the average amount of miles driven each year.

   $\sigma = 4688$ $\bar{x} = 17384$ $n = 500$ $c - level = .90$ $(17039, 17729)$

   I am 90% confident that the average number of miles driven per year is between 17039 and 17729 miles.

3. According to a poll of 100 randomly selected Americans, the average number of days spent on vacation is 5.6 with a standard deviation of 1.2. Calculate and interpret a 99% confidence interval for the average number of days Americans spend on vacation.

   $\sigma = 1.2$ $\bar{x} = 5.6$ $n = 100$ $c - level = .99$ $(5.2909, 5.9091)$

   I am 99% confident the average number of days Americans spent on vacation is between 5.2909 and 5.9091

4. A random sample of 400 Americans revealed the average amount of water consumed per day is 32.5 ounces, with a standard deviation of 10.6 ounces. Calculate and interpret a 90% confidence interval for the average number of ounces of water consumed per day by Americans.

   $\sigma = 10.6$ $\bar{x} = 32.5$ $n = 400$ $c - level = .90$ $(31.6282, 33.3718)$

   I am 90% confident that the average amount of water Americans consume per day is between 31.6282 and 33.3718 ounces.

Mixed Confidence Interval Problems

1. A sleeping pill was given to 300 randomly selected people, and the amount of time it took for the people to fall asleep was recorded. The average was 19.5 minutes with a standard deviation of 3.4 minutes. Use this information to calculate and interpret a 99% confidence interval for the average amount of time it should take all people to fall asleep after taking this pill, if the pill were to be released nationwide.

2. In a random sample of 600 people, 12 had an allergic reaction to an experimental drug. Calculate and interpret a 95% confidence interval for the proportion of all people who would have an allergic reaction to this drug.

3. A random sample of 350 people consumed an average of 2033 calories per day, with a standard deviation of 315 calories. Calculate and interpret a 99% confidence interval for the average number of calories consumed per day.

4. A random sample of 300 voters revealed that 53% are dissatisfied with a newly proposed law. Calculate and interpret a 90% confidence interval for the proportion of all voters who are dissatisfied with the newly proposed law.

Mixed Confidence Interval Problems

1. A sleeping pill was given to 300 randomly selected people, and the amount of time it took for the people to fall asleep was recorded. The average was 19.5 minutes with a standard deviation of 3.4 minutes. Use this information to calculate and interpret a 99% confidence interval for the average amount of time it should take all people to fall asleep after taking this pill, if the pill were to be released nationwide.

Means problem

$\sigma = 3.4$ $\bar{x} = 19.5$ $n = 300$ $c - level = .99$ **(18.994, 20.006)**

I am 99% confident that the average amount of time it will take all people to fall asleep after taking this pill is between 18.994 and 20.006 minutes.

2. In a random sample of 600 people, 12 had an allergic reaction to an experimental drug. Calculate and interpret a 95% confidence interval for the proportion of all people who would have an allergic reaction to this drug.

Proportions problem   x = 12   n = 600   c-level = .95   (.0088, .0312)

I am 95% confident that the proportion of all people who will have an allergic reaction to this drug is between .0088 and .0312.

3. A random sample of 350 people consumed an average of 2033 calories per day, with a standard deviation of 315 calories. Calculate and interpret a 99% confidence interval for the average number of calories consumed per day.

Means problem

$\sigma = 315$ $\bar{x} = 2033$ $n = 350$ $c - level = .99$ **(1989.6, 2076.4) I am 99% confident that the average number of calories consumed per day by all people is between 1989.6 and 2076.4 calories.**

4. A random sample of 300 voters revealed that 53% are dissatisfied with a newly proposed law. Calculate and interpret a 90% confidence interval for the proportion of all voters who are dissatisfied with the newly proposed law.

Proportions problem

x = .53 * 300 = 159   n = 300   c-level = .90   **(.4826, .5774)   I am 95% confident that the proportion of all people who are dissatisfied with a newly proposed law is between .4826and .5774.**

Mixed Confidence Interval Problems

1.   A bowler recorded her score for 58 randomly selected games; the average was 257.93 with a standard deviation of 15.85. Calculate and interpret a 95% confidence interval for her average score for all games.

2.   A random sample of 290 people revealed that 175 file their taxes in April. Calculate and interpret a 95% confidence interval for the proportion of all people who file their taxes in April.

3.   A local store owner printed a coupon in a paper. The following week, she noted that 38% of 200 randomly selected customers used the coupon. Calculate and interpret a 99% confidence interval for the proportion of all customers who used the coupon.

4.   A randomly selected group of 200 teachers revealed an average of 5.4 sick days per year, with a standard deviation of 1.2 days. Calculate and interpret a 90% confidence interval for the average number of sick days taken by teachers in a year.

Mixed Confidence Interval Problems

1. A bowler recorded her score for 58 randomly selected games; the average was 257.93 with a standard deviation of 15.85. Calculate and interpret a 95% confidence interval for her average score for all games. **Means problem**

$\sigma = 15.85$ $\bar{x} = 257.93$ $n = 58$ $c - level = .95$ (253.851, 262.009)

I am 99% confident that the average bowling score for all of her games is between 253.851 and 262.009.

2. A random sample of 290 people revealed that 175 file their taxes in April. Calculate and interpret a 95% confidence interval for the proportion of all people who file their taxes in April.

Proportions problem x = 175   n = 290   c-level = .95   (.5471, .6598)

I am 95% confident that the proportion of all people who file their taxes in April is between .5471 and .6598

3. A local store owner printed a coupon in a paper. The following week, she noted that 38% of 200 randomly selected customers used the coupon. Calculate and interpret a 99% confidence interval for the proportion of all customers who used the coupon.

Proportions problem

x = .38 * 200 = 76   n = 200   c-level = .99   (.2916, .4684)   **I am 99% confident that the proportion of all people who use the coupon is between .2916 and .4684**

4. A randomly selected group of 200 teachers revealed an average of 5.4 sick days per year, with a standard deviation of 1.2 days. Calculate and interpret a 90% confidence interval for the average number of sick days taken by teachers in a year. **Means problem**

σ=1.2   x =5.4   n=200   c-level= .90   (5.2604, 5.5396)

I am 99% confident that the average number of sick days per year for all teachers is between 5.2604 and 5.5396 days.

Hypothesis Testing on Proportions

1. A recent study showed that 34% of released criminals reoffend within six months. You believe the percentage is lower than that. In a random sample of 350 released criminals, 112 had reoffended within six months. Is this significant at the .05 level?

2. Somebody claimed that 75% of people get their news from social media. You believe the percentage is lower than that. After randomly selecting 300 people, you determined that 72% said they get their news from social media. Is this significant at the .01 level? In other words, did you prove your point?

Hypothesis Testing on Proportions

1. A recent study showed that 34% of released criminals reoffend within six months. You believe the percentage is lower than that. In a random sample of 350 release criminals, 112 had reoffended within six months. Is this significant at the .05 level? In other words, did you prove your point?

Null: p = .34   Alternative: p < .34

Calculator: Po = .34   x = 112   n = 350   Alternative: <

Z = -.7899   p-value = .2148   α = .0500   p-value falls to the right of α

Fail to reject the null. This is not significant. I didn't prove that the percentage is lower.

2. Somebody claimed that 75% of people get their news from social media. You believe the percentage is lower than that. After randomly selecting 300 people, you determined that 72% said they get their news from social media. Is this significant at the .01 level?

Null: p = .75   Alternative: p < .75

Calculator: Po = .75   x = .72*300 = 216   n = 300   Alternative: <

z = -1.2   p-value = .1151   α = .0100   p-value falls to the right of α

Fail to reject the null. This is not significant. I didn't prove that the percentage is lower.

3. A mayor claimed that 87% of the high school students in his town go on to college. You want to test to see if the real percentage is different. In a random sample of 400 recent graduates, 333 went to college. Test this hypothesis at the .05 level. Did you show the real percentage is different?

4. A phone company strives for their satisfaction rate to be more than 90%. In a recent survey of 230 randomly selected customers, 93.04% said they are satisfied. Does this show, at the .01 level, that more than 90% of the customers are satisfied?

3. A mayor claimed that 87% of the high school students in his town go on to college. You want to test to see if the real percentage is different. In a random sample of 400 recent graduates, 333 went to college. Test this hypothesis at the .05 level. Did you show the real percentage is different?

Null: p = .87   Alternative: p ≠ .87

Calculator: Po = .87   x = 339   n = 400   Alternative: ≠

z = -2.2301   p-value = .0257   α = .0500   p-value falls to the left of α

Reject the null. This is significant. I proved the actual percentage is different from 87%

4. A phone company strives for their satisfaction rate to be more than 90%. In a recent survey of 230 randomly selected customers, 93.04% said they are satisfied. Does this show, at the .01 level, that more than 90% of the customers are satisfied?

Null: p = .90   Alternative: p > .90

Calculator: Po = .90   x =.9304 * 230 = 214   n = 230   Alternative: >

z = 1.5386   p-value = .0620   α = .0100 p-value falls to the right of α

Fail to reject the null. This does not show that more than 90% of all customers are satisfied.

Hypothesis Testing on Means

1. A pill manufacturer claims that the average improvement in blood pressure is 14 points after patients take their medication. The standard deviation is known to be 5.1 points. You want to see if this test is correct or if the real average is different from that. A random sample of 500 patients showed an average improvement of 13.5 points. Is this significant at the .05 level? Does it appear that the average is 14?

2. A high school claims their average SAT score is 1683. You believe it is lower than that. A random sample of 100 students from that school had an average SAT score of 1651.3. The standard deviation of SAT scores is known to be 300. Does this show, at significance level .05, that the average is really less than 1683?

Hypothesis Testing on Means

1.   A pill manufacturer claims that the average improvement in blood pressure is 14 points after patients take their medication. The standard deviation is known to be 5.1 points. You want to see if this test is correct or if the real average is different from that. A random sample of 500 patients showed an average improvement of 13.5 points. Is this significant at the .05 level? Does it appear that the average is 14?

Null: $\mu = 14$ Alt: $\mu \neq 14$

Calculator: μ_0: 14 σ :5.1 $\bar{x}$: 13.5   n : 500   sign : $\neq$

z-score = -2.1922   p-value = .0284 α = .05

Reject the null. The evidence suggests that the true average is different from 14.

2.   A high school claims their average SAT score is 1683. You believe it is lower than that. A random sample of 100 students from that school had an average SAT score of 1651.3. The standard deviation of SAT scores is known to be 300. Does this show, at significance level .05, that the average is really less than 1683?

Null: $\mu = 1683$ Alt: $\mu < 1683$

Calculator: μ_0: 1683 σ :300 $\bar{x}$: 1651.3   n : 100   sign : <

z-score = -1.0567   p-value = .1453 α = .05

Fail to reject the null. The evidence does not suggest that the true average is below 1683.

3.   The average birth weight for newborn babies is 7.3 pounds with a standard deviation of 1.9 pounds. You wish to see if a particular disease causes women to have, on average, larger babies. A random sample of 200 babies born to women with this disease was collected, and their average birth weight was 7.7 pounds. Is this significant at the .02 level? Did you successfully prove the babies are larger?

4.   The average length of a commercial is claimed to be 30 seconds, with a standard deviation of 2.4 seconds. You believe the average length for the advertising company you work for is less than that. A random sample of 40 commercials produced by your company had an average length of 29.1 seconds. Is this significant at the .05 level? Did you show your company had a shorter average length?

3. The average birth weight for newborn babies is 7.3 pounds with a standard deviation of 1.9 pounds. You wish to see if a particular disease causes women to have, on average, larger babies. A random sample of 200 babies born to women with this disease was collected, and their average birth weight was 7.7 pounds. Is this significant at the .02 level? Did you successfully prove the babies are larger?

Null: $\mu = 7.3$　　　　　　　Alt: $\mu > 7.3$

Calculator:　μ_0: 7.3　　　σ :1.9　　　　$\bar{x}$: 7.7　　　　n : 200　　　sign : >

z-score = 2.9773　　　p-value = .0015　　　α = .02

Reject the null. The evidence suggests that the true average weight for these babies is more than 7.3 pounds.

4. The average length of a commercial is claimed to be 30 seconds, with a standard deviation of 2.4 seconds. You believe the average length for the advertising company you work for is less than that. A random sample of 40 commercials produced by your company had an average length of 29.1 seconds. Is this significant at the .05 level? Did you show your company had a shorter average length?

Null: $\mu = 30$　　　　　　　Alt: $\mu < 30$

Calculator:　μ_0 : 30　　　σ :2.4　　　　$\bar{x}$: 29.1　　　n : 40　　　sign : <

z-score = -2.3717　　　p-value = .0089　　　α = .05

Reject the null. The evidence suggests that the true average length for your company's commercials is less than 30 seconds.

Hypothesis Testing, Mixed Examples

1. A politician claims that 73% of people approve of how she is handling a particular crisis. You believe that claim is false, and the real percentage is lower than that. In a random sample of 200 people, 140 say they approve of her strategy. Is this significant at the .05 level? Did you show that the real percentage is lower?

2. An old study showed that the average size of crabs coming from the Chesapeake Bay is 4.56 inches across, with a standard deviation of .4 inches. You believe that, due to pollution, the average size is now smaller. A random sample of 50 crabs showed an average size of 4.43 inches. Is this significant at the .05 level? Did you show the crabs are, on average, now smaller?

Hypothesis Testing, Mixed Examples

1.   A politician claims that 73% of people approve of how she is handling a particular crisis. You believe that claim is false, and the real percentage is lower than that. In a random sample of 200 people, 140 say they approve of her strategy. Is this significant at the .05 level? Did you show that the real percentage is lower?

Proportions problem

Null: p = .73   Alternative: p < .73

Calculator: Po = .73   x =140   n = 200   Alternative: <

z = -.9556   p-value = .1696 α = .0500   p-value falls to the right of α

Fail to reject the null. This not significant. I failed to prove the real percentage is lower than 73%

2.   An old study showed that the average size of crabs coming from the Chesapeake Bay is 4.56 inches across, with a standard deviation of .4 inches. You believe that, due to pollution, the average size is now smaller. A random sample of 50 crabs showed an average size of 4.43 inches. Is this significant at the .05 level? Did you show the crabs are, on average, now smaller?

Means problem

Null: $\mu = 4.56$ **Alt: $\mu < 4.56$**

Calculator: μ_0 : 4.56 σ :0.4 $\bar{x}$: 4.43   n : 50   sign : <

z-score = -2.2981   p-value = .0108 α = .05

Reject the null. The evidence suggests that the true average length of the crabs is now smaller. (It is significant)

3. Once diagnosed with Cushing's Disease, the average amount of time a dog lives is 3.1 years with a standard deviation of 0.6 years. You have a new medication that you believe will lengthen the average amount of time these dogs can survive. A random sample of 26 dogs who had just been diagnosed with Cushing's Disease took your medication, and they lived an average of 3.4 years. Does this show at significance level .01 that your medication makes the average survival longer?

4. A cell phone carrier claims that 43% of people in your county use their service. You believe it is less than that. A random sample of 200 people revealed that 41.5% of them used the service. Is that significant at the .05 level? Did you show the actual percentage is lower?

3.   Once diagnosed with Cushing's Disease, the average amount of time a dog lives is 3.1 years with a standard deviation of 0.6 years. You have a new medication that you believe will lengthen the average amount of time these dogs can survive. A random sample of 26 dogs who had just been diagnosed with Cushing's Disease took your medication, and they lived an average of 3.4 years. Does this show at significance level .01 that your medication makes the average survival longer?

Means problem

Null: $\mu = 3.1$ **Alt:** $\mu > 3.1$

Calculator: $\mu_0 : 3.1$ $\sigma : 0.6$ $\bar{x} : 3.4$ **n : 26**   **sign : >**

z-score = 2.5495   **p-value = .0054** $\alpha = .01$

Reject the null. The evidence suggests that the true average survival length will be longer on this medication. (It is significant)

4.   A cell phone carrier claims that 43% of people in your county use their service. You believe it is less than that. A random sample of 200 people revealed that 41.5% of them used the service. Is that significant at the .05 level? Did you show the actual percentage is lower?

Proportions problem

Null: p = .43   Alternative: p < .43

Calculator: Po = 43   x = .415 (200) = **83**   n = 200   Alternative: <

z = .4285   p-value = .3341 α = .0500   p-value falls to the right of α

Fail to reject the null. This not significant. I failed to prove the real percentage is lower than 43%

Confidence Interval Z* Values

Confidence Level	Z*
68%	1
90%	1.645
95%	1.960 (or 2)
99%	2.576
99.7%	3

Z table

z	.00	.01	.02	.03	.04	.05	.06	.07	.08	.09
-3.4	.0003	.0003	.0003	.0003	.0003	.0003	.0003	.0003	.0003	.0002
-3.3	**.0005**	**.0005**	**.0005**	**.0004**	**.0004**	**.0004**	**.0004**	**.0004**	**.0004**	**.0003**
-3.2	.0007	.0007	.0006	.0006	.0006	.0006	.0006	.0005	.0005	.0005
-3.1	**.0010**	**.0009**	**.0009**	**.0009**	**.0008**	**.0008**	**.0008**	**.0008**	**.0007**	**.0007**
-3.0	.0013	.0013	.0013	.0012	.0012	.0011	.0011	.0011	.0010	.0010
-2.9	**.0019**	**.0018**	**.0018**	**.0017**	**.0016**	**.0016**	**.0015**	**.0015**	**.0014**	**.0014**
-2.8	.0026	.0025	.0024	.0023	.0023	.0022	.0021	.0021	.0020	.0019
-2.7	**.0035**	**.0034**	**.0033**	**.0032**	**.0031**	**.0030**	**.0029**	**.0028**	**.0027**	**.0026**
-2.6	.0047	.0045	.0044	.0043	.0041	.0040	.0039	.0038	.0037	.0036
-2.5	**.0062**	**.0060**	**.0059**	**.0057**	**.0055**	**.0054**	**.0052**	**.0051**	**.0049**	**.0048**
-2.4	.0082	.0080	.0078	.0075	.0073	.0071	.0069	.0068	.0066	.0064
-2.3	**.0107**	**.0104**	**.0102**	**.0099**	**.0096**	**.0094**	**.0091**	**.0089**	**.0087**	**.0084**
-2.2	.0139	.0136	.0132	.0129	.0125	.0122	.0119	.0116	.0113	.0110
-2.1	**.0179**	**.0174**	**.0170**	**.0166**	**.0162**	**.0158**	**.0154**	**.0150**	**.0146**	**.0143**
-2.0	.0228	.0222	.0217	.0212	.0207	.0202	.0197	.0192	.0188	.0183
-1.9	**.0287**	**.0281**	**.0274**	**.0268**	**.0262**	**.0256**	**.0250**	**.0244**	**.0239**	**.0233**
-1.8	.0359	.0351	.0344	.0336	.0329	.0322	.0314	.0307	.0301	.0294
-1.7	**.0446**	**.0436**	**.0427**	**.0418**	**.0409**	**.0401**	**.0392**	**.0384**	**.0375**	**.0367**
-1.6	.0548	.0537	.0526	.0516	.0505	.0495	.0485	.0475	.0465	.0455
-1.5	**.0668**	**.0655**	**.0643**	**.0630**	**.0618**	**.0606**	**.0594**	**.0582**	**.0571**	**.0559**
-1.4	.0808	.0793	.0778	.0764	.0749	.0735	.0721	.0708	.0694	.0681
-1.3	**.0968**	**.0951**	**.0934**	**.0918**	**.0901**	**.0885**	**.0869**	**.0853**	**.0838**	**.0823**
-1.2	.1151	.1131	.1112	.1093	.1075	.1056	.1038	.1020	.1003	.0985
-1.1	**.1357**	**.1335**	**.1314**	**.1292**	**.1271**	**.1251**	**.1230**	**.1210**	**.1190**	**.1170**
-1.0	.1587	.1562	.1539	.1515	.1492	.1469	.1446	.1423	.1401	.1379
-0.9	**.1841**	**.1814**	**.1788**	**.1762**	**.1736**	**.1711**	**.1685**	**.1660**	**.1635**	**.1611**
-0.8	.2119	.2090	.2061	.2033	.2005	.1977	.1949	.1922	.1984	.1867
-0.7	**.2420**	**.2389**	**.2358**	**.2327**	**.2296**	**.2266**	**.2236**	**.2206**	**.2177**	**.2148**
-0.6	.2743	.2709	.2676	.2643	.2611	.2578	.2546	.2514	.2483	.2451
-0.5	**.3085**	**.3050**	**.3015**	**.2981**	**.2946**	**.2912**	**.2877**	**.2843**	**.2810**	**.2776**
-0.4	.3446	.3409	.3372	.3336	.3300	.3264	.3228	.3192	.3156	.3121
-0.3	**.3821**	**.3783**	**.3745**	**.3707**	**.3669**	**.3632**	**.3594**	**.3557**	**.3520**	**.3483**
-0.2	.4207	.4168	.4129	.4090	.4052	.4013	.3974	.3936	.3897	.3859
-0.1	**.4602**	**.4562**	**.4522**	**.4483**	**.4443**	**.4404**	**.4364**	**.4325**	**.4286**	**.4247**
-0.0	.5000	.4960	.4920	.4880	.4840	.4801	.4761	.4721	.4681	.4641

z	.00	.01	.02	.03	.04	.05	.06	.07	.08	.09
0.0	.5000	.5040	.5080	.5120	.5160	.5199	.5329	.5279	.5319	.5359
0.1	**.5398**	**.5438**	**.5478**	**.5517**	**.5557**	**.5596**	**.5636**	**.5675**	**.5714**	**.5753**
0.2	.5793	.5832	.5871	.5910	.5948	.5987	.6026	.6064	.6103	.6141
0.3	**.6179**	**.6217**	**.6255**	**.6293**	**.6331**	**.6368**	**.6406**	**.6443**	**.6480**	**.6517**
0.4	.6554	.6591	.6628	.6664	.6700	.6736	.6772	.6808	.6844	.6879
0.5	**.6915**	**.6950**	**.6985**	**.7019**	**.7054**	**.7088**	**.7123**	**.7157**	**.7190**	**.7224**
0.6	.7257	.7291	.7324	.7357	.7389	.7422	.7454	.7486	.7517	.7549
0.7	**.7580**	**.7611**	**.7642**	**.7673**	**.7704**	**.7734**	**.7764**	**.7794**	**.7823**	**.7852**
0.8	.7881	.7910	.7939	.7967	.7995	.8023	.8051	8078	.8106	.8133
0.9	**.8159**	**.8186**	**.8212**	**.8238**	**.8264**	**.8289**	**.8315**	**.8340**	**.8365**	**.8389**
1.0	.8413	.8438	.8461	.8485	.8508	.8531	.8554	.8577	.8599	.8621
1.1	**.8643**	**.8665**	**.8686**	**.8708**	**.8729**	**.8749**	**.8770**	**.8790**	**.8810**	**.8830**
1.2	.8849	.8869	.8888	.8907	.8925	.8944	.8962	.8980	.8997	.9015
1.3	**.9032**	**.9049**	**.9066**	**.9082**	**.9099**	**.9115**	**.9131**	**.9147**	**.9162**	**.9177**
1.4	.9192	.9207	.9222	.9236	.9251	.9265	.9279	.9292	.9306	.9319
1.5	**.9332**	**.9345**	**.9357**	**.9370**	**.9382**	**.9394**	**.9406**	**.9418**	**.9429**	**.9441**
1.6	.9452	.9463	.9474	.9484	.9495	.9505	.9515	.9525	.9535	.9545
1.7	**.9554**	**.9564**	**.9573**	**.9582**	**.9591**	**.9599**	**.9608**	**.9616**	**.9625**	**.9633**
1.8	.9641	.9649	.9656	.9664	.9671	.9678	.9686	.9693	.9699	.9706
1.9	**.9713**	**.9719**	**.9726**	**.9732**	**.9738**	**.9744**	**.9750**	**.9756**	**.9761**	**.9767**
2.0	.9772	.9778	.9783	.9788	.9793	.9798	.9803	.9808	.9812	.9817
2.1	**.9821**	**.9826**	**.9830**	**.9834**	**.9838**	**.9842**	**.9846**	**.9850**	**.9854**	**.9857**
2.2	.9861	.9864	.9868	.9871	.9875	.9878	.9881	.9884	.9887	.9890
2.3	**.9893**	**.9896**	**.9898**	**.9901**	**.9904**	**.9906**	**.9909**	**.9911**	**.9913**	**.9916**
2.4	.9918	.9920	.9922	.9925	.9927	.9929	.9931	.9932	.9934	.9936
2.5	**.9938**	**.9940**	**.9941**	**.9943**	**.9945**	**.9946**	**.9948**	**.9949**	**.9951**	**.9952**
2.6	.9953	.9955	.9956	.9957	.9959	.9960	.9961	.9962	.9963	.9964
2.7	**.9965**	**.9966**	**.9967**	**.9968**	**.9969**	**.9970**	**.9971**	**.9972**	**.9973**	**.9974**
2.8	.9974	.9975	.9976	.9977	.9977	.9978	.9979	.9979	.9980	.9981
2.9	**.9981**	**.9982**	**.9982**	**.9983**	**.9984**	**.9984**	**.9985**	**.9985**	**.9986**	**.9986**
3.0	.9987	.9987	.9987	.9988	.9988	.9989	.9989	.9989	.9990	.9990
3.1	**.9990**	**.9991**	**.9991**	**.9991**	**.9992**	**.9992**	**.9992**	**.9992**	**.9993**	**.9993**
3.2	.9993	.9993	.9994	.9994	.9994	.9994	.9994	.9995	.9995	.9995
3.3	**.9995**	**.9995**	**.9995**	**.9996**	**.9996**	**.9996**	**.9996**	**.9996**	**.9996**	**.9997**
3.4	.9997	.9997	.9997	.9997	.9997	.9997	.9997	.9997	.9997	.9998

?